What on Earth Evolved?
...in Brief

CHRISTOPHER LLOYD read history at Cambridge, where he graduated with a double first, before becoming technology correspondent for the *Sunday Times*. In 1994 he won the Texaco award for the Science Journalist of the Year. After leaving journalism, he ran a number of Internet and educational publishing businesses. In 2006 he decided to home-school his two daughters, and the inspiration to write his first book, *What on Earth Happened?*, published in 2008, came during a four-month tour of Europe with his family.

Picture credits

The Advertising Archives: plates 10, 11 (top), 12 (top). **The Art Archive:** plates 4 (bottom), 6 (top), 7 (top), 9, 12 (bottom), 14, 15. **The Bridgeman Art Library:** plates 6 (bottom), 7 (bottom). **Corbis Images:** plates 3 (bottom), 5 (bottom), 11 (bottom), 13 (top). **Getty images:** plate 4 (top left). **Natural History Museum:** plate 5 (top). **Science Photo Library:** plates 1, 2, 3 (top), 4 (top right), 13 (bottom), 16. **The Wellcome institute:** plate 8.

First published in Great Britain 2009
This paperback edition published 2010

Bloomsbury Publishing Plc, 36 Soho Square, London W1D 3QY

A CIP catalogue record for this book is available from the British Library

ISBN 9781408802892

10 9 8 7 6 5 4 3 2 1

Typeset by Hewer Text UK Ltd, Edinburgh
Printed in Great Britain by Clays Ltd, St Ives plc

Mixed Sources
Product group from well-managed
forests and other controlled sources
www.fsc.org Cert no. SGS-COC-2061
© 1996 Forest Stewardship Council

www.bloomsbury.com/christopherlloyd

Comments and suggestions about this book can be posted on the discussion forum at
www.whatonearthevolved.com. All contributions and feedback are most welcome.

What on Earth Evolved? ... in Brief

100 species that have changed the world

Christopher Lloyd

Illustrations by Andy Forshaw

BLOOMSBURY

LONDON · BERLIN · NEW YORK

Before Humans

On the impact of species that evolved in the wild

5. On Pioneers of the Land

How mutual collaboration between trailblazing species helped life colonize the Earth's barren landscape. **75**

6. On Fish that Came Ashore

How the descendants of bony fish clambered ashore and were jerry-wrenched into a diverse range of forms. **105**

7. On Biodiversity

How beauty, collaboration, deception, parasitism and vice wove terrestrial life into a rich carpet of countless species. **129**

8. On the Rise of Reason

How the mental skills of mammals gave rise to a new evolutionary force that turned monkeys into men. **167**

After Humans page 199

On the impact of species that thrived in the presence of modern mankind

13. On Beauty

How good looks, strong smells and powerful flavours in some species have proved irresistible to humans. **353**

14. On Rivalry

How certain species have thrived with the rise of human civilizations, often despite attempts to keep them at bay. **395**

The Ladder of Life 427

The top 100 species ranked in order of overall impact

Postscript 441

Thirty species that nearly made it!

Introduction

What is life? What forces are in control? Why have creatures evolved as they are? Where does humanity fit in?

UNLIKE CHARLES DARWIN'S theory published 150 years ago, this book is not chiefly concerned with the 'origin of species', but with the influences and impacts that living things have had on the path of evolution, on each other and on our mutual environment, planet Earth.

This is an abridged verion of *What on Earth Evolved?*, roughly half the length of the original hardback. Although it excludes a number of narratives, which put the impact of each group of species into their overall evolutionary context, I hope you agree that the individual biographies of those 100 extraordinary living things still speak for themselves.

The first part surveys the mechanics of evolution before mankind, from the earliest replicating molecules to the rapid rise of mammals following the death of the dinosaurs, profiles fifty of the most successful life-forms (from now on, informally described as 'species') that emerged through *natural selection* up to the time when modern humans first trod the Earth.

What on Earth Evolved? . . . in Brief

The second part assesses how from *c.* 12,000 years ago humans introduced new evolutionary forces through the cultivation of plant and animal breeds to benefit human societies. The success of many creatures then came to depend on their capacity to impress and accommodate the needs of humans, which often replaced the traditional struggle for survival in the wild. What impact did these evolutionary forces have on the non-human world and on the direction of evolution itself? The second part profiles fifty of the most successful species that have developed through *artificial selection* up to the present day.

Finally, part three ranks the impact of the hundred chosen species – from the dawn of evolution to the present day – into a table of influence based on a set of simple, but arbitrary, criteria. Its intention is simply to stimulate thought, provoke comment and offer an alternative, sideways glance through the history of life.

The central purpose of this book is to cultivate a richer understanding of all history – not just as chronology but as seen through the lens of the natural world itself. Its aim is not only to inform and entertain but, I hope, to stimulate debate about the place of mankind in nature and its pivotal relationship with non-human life and the Earth itself.

Christopher Lloyd
June 2010

Before Humans

From four billion to 12,000 years before present . . .

On the impact of species that evolved in the wild

1

On Viruses

How loose strands of genetic code swarmed across the early Earth, inserting themselves into all forms of life – past, present and future.

Influenza

FAMILY: ORTHOMYXOVIRIDAE
SPECIES: INFLUENZAVIRUS A
RANK: 9

One of humanity's biggest ever killers

IN 1798 ENGLISH economist Thomas Malthus predicted the imminent decimation of human populations by *'sickly seasons, epidemics, pestilence and plague'* that *'advance in terrific array, and sweep off their thousands and tens of thousands . . .'* This alarming prophecy was founded on the rapid rise of the human population. Not long after Malthus wrote his famous essay, the number of humans inhabiting the Earth rose past the one billion mark. Just over 200 years later, the human population stands at nearly seven billion, with approximately 211,000 extra bodies added every day. Nothing could be better for viruses whose single concern is to copy their codes into as many other creatures (hosts) as possible. All living organisms from bacteria, plants and fungi to animals, fish and humans are vulnerable to infection by viruses – so great is their variety.

Influenza viruses infect a wide range of creatures including humans, pigs, birds, seals, mosquitoes, salmon and sea lice. At some point in the last few thousand years a variant of this virus family skipped across the species barrier and mutated, allowing it to spread from human to human. Its job was made much easier by the closer proximity between humans and animals that resulted from the rise of animal domestication and farming, beginning about 10,000 years ago (see page 201). The influenza virus spreads by causing its host to erupt into severe bouts of coughing and sneezing. In dense human populations copies of the influenza virus are easily spluttered from person to person in airborne droplets. The human immune system is usually able to overcome these infections within two or three weeks. Old and young people are at greatest risk of severe illness. On average one person dies out of every one thousand infected.

Since Malthus wrote his essay, more than *one hundred million people* worldwide are thought to have died from various strains of flu. By far the worst outbreak came between 1918 and 1920 when between fifty million and one hundred million people fell victim. This was more than double the number of soldiers slaughtered in World War I (1914–18), which ended just as the influenza outbreak got under way. The epidemic, which struck first in the United States, had a truly global impact. It spread throughout Europe and Asia, as far north as the Eskimo populations of the Arctic and west to the remote Pacific islands. Yet it came to be called 'Spanish flu' mainly because newspaper reports about its progress were least censored in Spain (which was not involved in the war effort and therefore had no ministry of propaganda) giving rise to the mistaken impression that the disease had Spanish origins.

A new strain of bird flu (H5N1) evolved in Asia between 1999 and 2002; however, it has had a limited effect on humans because this strain can be transmitted only from birds to humans, but not human to human, thereby restricting the risk of infection to those people who live or work in close proximity to birds. Nevertheless, an outbreak of another influenza variant (H1N1), originating in pigs, has since migrated throughout the world. Its easy spread from human to human caused this strain to be officially declared a global pandemic by the World Health Organization in the summer of 2009. International efforts to contain the spread of this so-called 'swine flu' were a conspicuous failure.

It is a worrying sign. Viruses have an extraordinary capacity for genetic diversity and change. The copying process used by influenza (which is based on RNA chemical replicators) makes errors far more often than the more sophisticated DNA genetic copying processes used in plant and animal cells. More errors mean more mutations, leading to the potential evolution of highly infectious lethal strains that can easily pass from human to human, turning each new victim into a deadly carrier of the disease. Mass immunization may or may not be effective, since precise counter-measures cannot be designed until it is known exactly how the virus has mutated – knowledge that can be gathered only *after* a new strain of the virus has struck. Mass production of a suitable vaccine would take at least three months, by which time mutant strains of such an influenza virus could be wiping out tens if not hundreds of millions of people.

Malthus may have got his timing wrong, but in our overcrowded modern world and with highly infectious diseases like influenza, his prediction is as pertinent today as ever.

HIV/AIDS

> **FAMILY: RETROVIRIDAE**
> **SPECIES: HUMAN IMMUNODEFICIENCY VIRUS 1 & 2**
> **RANK: 77**

*A most perplexing and fast-spreading disease
with no obvious cure*

PEOPLE WHO LIVE in and around Cape Town lead desperate lives. Their sorry story, which is repeated for millions more who live in other southern African cities, began thirty years ago when a devastating viral infection skipped across the species barrier from monkeys to humans. No one quite knows exactly how or when the HIV virus claimed its first human victim but it is thought that the incurable illness that stems from its infection, called AIDS, has so far claimed more than twenty-five million lives in twenty-five years, with at least another sixty million infected.

HIV is today's best known but least understood virus. As a member of the retrovirus family its ingenious tactics for splicing its own genetic code into the genes of its human hosts are the product of billions of years of evolution. That's why, despite enormous advances in developing a vaccine for many of the most devastating human diseases, there is still no cure for HIV. Nor is there likely to be one any time soon.

Viruses like HIV replicate so fast and make so many copying errors in the process that they are usually able to outwit vaccines. After infecting its host through bodily fluids such as blood, semen, saliva or breast milk, HIV mutates into so many different forms that the human immune system can't keep up. In many people the HIV

7

infection can be contained by the body's defences for only about one year – although this can be greatly extended with the help of modern antiviral drugs. Eventually, however, the virus comes up with a new variant that the body's defences cannot circumvent. It then duplicates madly, wrecking what's left of its human host's disease-fighting systems. Death arises from other normally benign infections, such as the common cold, because the body has become too weak to ward them off.

A few people are naturally immune to HIV. A recent study has tried to explain why about 10 per cent of the European population benefit from a special type of mutation that is able to combat the spread of HIV. Research indicates that this mutation is present in human populations that suffered outbreaks of bubonic plague between c. AD 1000 and 1800. Viral haemorrhagic fever is first thought to have appeared about 2,500 years ago but infections became pandemic in Europe only after AD 1000, peaking with the Black Death in the mid-fourteenth century. Unfortunately, today's human population in Sub-Saharan Africa has no history of medieval bubonic plague. This infection has therefore gained its strongest grip in a place where natural human immunity is weakest.

Experts also believe that for billions of years viruses like HIV may have been instrumental in triggering the appearance of new species. It is interesting to speculate on what would happen to human populations in Africa in the absence of modern social, educational or technological intervention. Eventually the entire continent would become infected with HIV and all except for the few people with natural immunity would perish. It would be left to these immune survivors to repopulate the continent. But the new population would differ from the old one in genetic makeup by having a benign HIV signature in their genes, a signature that would effectively have been added to the ranks of these people's 'junk' DNA. Since human populations on other continents would not be able to breed with surviving African humans without themselves becoming infected, at least two isolated populations of humans would emerge – those with HIV in Africa and those without HIV in the rest of the world. The inability to interbreed between these populations means they would eventually 'evolve' into different human species.

Only about 1 per cent of African adults have so far been tested for HIV, so true infection rates are still largely unknown, making its future impact impossible to predict. In the absence of a vaccine or other cure, the disease continues to accelerate, devastating the social and economic lives of millions, the majority of whom live in the poorest parts of Africa and India. In 2007 an estimated 2.1 million humans died of AIDS, 330,000 of them children under the age of fifteen. South Africa has more than one million orphaned children, themselves HIV infected, thanks to the passage of the virus from mother to child via breast milk.

Meanwhile, in neighbouring Botswana average human life expectancy has plummeted from sixty-five years in 1988 to thirty-five in 2007. Despite advances in modern medicine, the impact of HIV infection on Sub-Saharan Africa today puts it squarely alongside history's most devastating viral pandemics: avian influenza (see page 5) and smallpox (see page 13).

Potyvirus

FAMILY: POTYVIRIDAE
SPECIES: POTYVIRUS SP.
RANK: 100

How a plant virus caused the first stock market crash

PLANTS ARE JUST as susceptible to infection by viruses as humans, animals, fungi or bacteria. Less well known is how one virus had a significant impact on the birth of botany, horticulture and capitalism in the history of early-modern Europe.

Potyviridae is the largest of the thirty-four known plant virus families. It comprises about 180 members that together account for 30 per cent of all recorded plant viruses. These parasites are not experts in sneakily copying their own genetic code into the DNA of their hosts, nor do they spread by direct contact between one plant and another. Instead, these viruses employ the same technique for their reproduction as many plants use for pollination – they hitch a lift from a flying passer-by. The potyvirus uses the transportation of a species of insect called the green peach aphid (*Myzus persicae*). Scuttling from plant to plant, these small creatures have spread potyviruses all over the world. They ride on the back of winds that blow them hundreds of miles across land and sea, and there is not a single region of the world that is not now affected by the viral infection they carry. Plants most at risk from damage include potatoes, turnips and plums, and in most cases, once infected, their fruit or tubular production becomes seriously impaired. However, Potyviridae's biggest impact on human history comes from one variant that ably demonstrates how viral infection can sometimes turn out to be highly beneficial for the survival prospects of its host species.

Tulips originated in central Asia. They grow naturally in the mountainous regions of Turkey, Iran and Afghanistan. Long admired by Islamic sultans, these flowers were brought to Europe in the late sixteenth century by a Flemish doctor called Carolus Clusius (1526–1609) who was also responsible for running the imperial medicinal garden in Vienna. In 1593, when he moved to Leiden, southern Holland, he took with him the bulbs of some especially attractive tulips. As if by magic, some of the plants flowered with a highly distinctive flaming pattern that unfurled across their petals like marbled paper. No one knew why most tulips stayed a single, monotonous shade while a precious few broke into such a feast of whimsical streaks and stripes. The powerful allure of these exotic varieties among Dutch traders meant that, for a while, tulips became the most sought-after, valuable commodity money could buy.

So feverish did the Dutch appetite for tulips become that contracts for flowers due to be harvested in the autumn of 1635 were sold for more than the price of a house. The most celebrated was the vivid variety *Semper Augustus*, painted in all its majesty by famous Dutch artists of the day. A dozen or so samples owned by one tulip collector, Dr Adriaen Pauw, accounted for almost the entire world's supply, so rare was this stunning beauty. Despite numerous offers throughout the 1620s, Pauw refused to sell his collection at any price. When a single *Semper Augustus* bulb finally came up for auction, it sold for a record 6,000 florins (about £1 million today).

Prices like these were driven by scarcity. Even though tulip flowers are capable of producing up to 200 seeds a year, they hardly ever pass on their striking looks to their offspring (for an explanation see apple, page 358). The only way to ensure a further crop of similarly patterned plants is to take off-cuts from the original plant (a form of cloning) and, annoyingly for Dutch growers, it seemed that the more beautiful the tulip, the weaker and less numerous were its bulbs from which off-cuts could be cultivated, partly explaining why prices shot through the roof. It wasn't until the invention of the electron microscope in the 1920s that scientists discovered that a virus prevented the production of certain petal pigments, causing the colour to 'break'. Only then was the mystery of why some tulips had such stunning looks resolved, and the link between virus, aphid and tulip finally pieced together.

The tulip speculators' bubble finally burst in the spring of 1637. Gamblers who had promised to pay thousands of florins for that year's harvest were left with nothing and those who had bought tulips the previous year for vast sums saw their fortunes evaporate overnight. Some historians claim it took years for the Dutch economy to recover from 'Tulip mania'. Others say its impact has been over-hyped. From the 1920s onwards the virus was systematically eradicated by Dutch farmers anxious to increase the yields of their tulip crop. By producing hybrids of other species, similar visual effects to those originally produced by the virus could be cultivated for less cost, time and effort.

Yet it is a fascinating and significant historical fact that a plant virus was instrumental in creating the first ever speculative financial crash. Tulip mania was an early forerunner of the South Sea Bubble, and more recently the dot-com boom and sub-prime mortgage crisis. More significantly, though, this virus-induced episode firmly established the Netherlands as a world centre of botany and horticulture and shows, however inadvertently, that a virus can sometimes help the survival prospects of its host species.

Smallpox

> **FAMILY: POXVIRIDAE**
> **SPECIES: VARIOLA VERA**
> **RANK: 63**

Humanity's most lethal viral killer

JANET PARKER WAS an exceptionally unlucky woman. Her tragic death on 11 September 1978 marked the passing of a long and inauspicious era in the history of human disease. Only six weeks earlier, she had been infected by a rogue strain of the smallpox virus *Variola major* that had escaped through an air vent from an unregulated research laboratory beneath her photographic studio in a university campus building in Birmingham, England. Parker, aged forty, was the last victim of the smallpox virus. Within only two years of her death, the World Health Organization officially declared the disease extinct in nature – the first and only case of a natural infection being totally eradicated by collective human endeavour.

Despite its recent extinction, smallpox has been one of humanity's most lethal viral killers. The disease passes easily through the air between one human and another whenever they are about two metres or less apart. Typically, about 30 per cent of those infected subsequently die, although in some outbreaks the rate increases to nearer 100 per cent. Cumulatively this disease has been responsible for the deaths of hundreds of millions, if not billions, of people.

Variola infects only humans. It began as a mutant strain that originated in another member of the Poxviridae family, cowpox (*Vaccinia*). Cowpox and monkeypox are non-lethal infections that can affect both humans and other animals. At some point after humans started keeping farm animals, about 10,000 years ago, a variant jumped the species barrier from cows to humans and evolved into a lethal new strain.

Poxviridae do not hijack the nucleus of a host's cell to replicate. Instead, they use their own double-stranded DNA as copying machinery, only helping themselves to their host cell's cytoplasm as a source of body-building protein. This unique behaviour has led some virologists to hypothesize that at a very early stage in the evolution of life on Earth a pox-like virus invaded simple bacterial cells (Prokaryotes), seeding them with a nucleus of DNA. If so, the effect that the ancient ancestors of the smallpox virus had on life on Earth were monumental. Their vital impact may therefore have been to provide the essential DNA genetic copying machinery that has made all forms of higher life possible.

Death by smallpox was extremely unpleasant. Signs of trouble emerged about twelve days after infection. Huge numbers of infected viral particles then spread through the bloodstream to all parts of the victim's body. High fever, muscle pain, headaches and vomiting gave way to a pernicious rash, usually inside the mouth, on the tongue, down the throat and all over the face. Rupturing blisters oozed copious doses of viral infection into the saliva. Two days later the rash spread all over the body, usually bursting out into large leaking pustules that eventually scabbed over. If the disease itself failed to kill its host, other infections such as bronchitis or pneumonia soon took over. The most fortunate survivors – like Queen Elizabeth I of England – were usually left with lifelong scars (in her case covered up with thick dollops of white lead makeup); others went permanently blind.

Understanding the enormous impact that this disease has had on human society is only possible where good historical records survive. One of the most ancient accounts is from the Greek historian Thucydides, who described a 'Plague of Athens' that decimated the city's population in 430 BC. From his account it seems almost certain that a smallpox epidemic contributed significantly to the Spartan victory in the famous Peloponnesian Wars (431–404 BC).

By 250 BC clear evidence emerges of smallpox infection in China, possibly spread by central Asian Huns. Frequent outbreaks followed in succeeding centuries and by the sixth century it had spread from Korea to Japan via Buddhist monks. Urbanization in Japan from 710 (with the building of the imperial capital of Nara) helped the disease spread, where it wiped out the ruling Fujiwara family and, from c. 735, became endemic (permanent in the society). One result was that Japan's Buddhist religious fervour grew more intense in a desperate attempt to relieve the empire of this appalling pestilence.

In Africa the disease long plagued the dynasties of ancient pharaohs, claiming the life of Rameses V in 1157 BC, whose mummy has recently been examined revealing a rash of raised pustules. It spread from there to the Middle East via the Hittite empire in Turkey. In c. AD 550 a mostly Christian army from Ethiopia invaded Yemen, reaching as far as Mecca in AD 570 – the year of the birth of the Muslim Prophet Mohammed. Then smallpox struck. The vastly outnumbered but infection-free Arabs were easily able to fight off the diseased Christians in a conflict known as the Elephant War. The virus travelled with Muslim warriors who pushed on into Spain, through the Middle East and even across the hot, dusty Saharan deserts into the plains of West Africa.

Smallpox had its greatest impact several hundred years later, carried by sixteenth-century newcomers to the Americas – the Spanish conquistadors. They were soon followed by West African slaves, shipped across the Atlantic to work in the plantations of the New World. Both waves of immigrants brought smallpox to the Americas, virtually annihilating the Native American populations who had no immunity to the disease because traditionally none had lived in close proximity to farm animals. One hundred years after the fall of the native Central and South American empires, the disease struck North America, this time spread by French settlers in Nova

Scotia. An epidemic starting in 1617 wiped out much of the Native American population along the east coast within just a few years. By the time the Pilgrim Fathers arrived from England on the *Mayflower* in 1621 nine tenths of the indigenous east coast population had died. Not that the European setters were totally immune; about twenty of those on the *Mayflower* also succumbed to the disease.

The first glimmer of a human fightback came in 1717, largely thanks to an enlightened English feminist. While living in Turkey, Lady Mary Montagu, wife of the British ambassador to Istanbul, wrote a series of letters describing life in and around the court of the Ottoman emperor. In one of them she described a method of 'inoculation' against smallpox. It is thought to have been pioneered in ancient India under the auspices of Hindu mystics who had their own goddess of smallpox, *Shitala Mata*. The process involved some powdered scab taken from a previously infected person who had recovered from the less virulent form of the disease, *Variola minor*. This was inserted into a small incision made in the skin of an uninfected person, provoking a mild form of the disease. Usually the process provided lifelong immunity, although it wasn't without risk. Lady Montagu was determined to lead the way and in 1721 she had both her children inoculated, the first such procedure ever performed in England.

While inoculation provided some protection from the disease, it wasn't until 1796 that a genuinely safe process was discovered by which immunity could be conferred without risk to doctor or patient. By injecting pus from a cowpox blister into a healthy human, Edward Jenner, a doctor from Gloucestershire, England, found a new method of immunization. He called it vaccination (from the Latin *vacca* for cow). Cowpox causes a mild and temporary skin infection in humans that the body's immune system has no problem controlling. If someone who has already had cowpox is later infected by smallpox, the body can successfully attack the disease because the viruses are so similar that it is tricked into thinking that the smallpox infection is a recurrence of cowpox. Immunity is almost always assured.

About 175 years passed between Jenner's discovery and the death of Janet Parker. In that short period, hundreds of millions of people continued to die from the disease, until its eventual eradication, certified in 1980. Today only two laboratories in the world are officially permitted to keep live samples of the virus for research in case of future outbreak – although to what extent other samples are held in secret by countries that were once part of the former Soviet Union remains unclear.

2

On Simple Cells

How versatile single living cells established new patterns of evolutionary behaviour, filling every available niche with life.

Cyanobacteria

FAMILY: SYNECHOCOCCACEA
SPECIES: PROCHLOROCOCCUS MARINUS
RANK: 3

*A photosynthesizing bacterium that provided
the essential breath of life*

IMAGINE A WORLD with no plants, trees or animals. The land would be a barren blend of rock and silt, the seas would be bereft of fish and the air silent except for the wind. This is a world devoid of a vital ingredient on which almost all forms of 'higher' life – fish, birds, insects, plants and animals – depend. It is a world without oxygen.

Until about fifty years ago no one imagined our planet Earth as a place without oxygen because throughout recorded human history (about 10,000 years) the amount of oxygen in the air has remained constant at a steady 21 per cent. Now we know differently. The Earth's first atmosphere probably contained only trace supplies of oxygen that quickly disappeared from the atmosphere after reacting with other elements, such as iron, to form rocky ores of iron oxide. Over the last few thousands years these rocks have been extensively mined by humans to extract the iron for crafting tools, making weapons and building shelters.

Plants, trees and animals all owe their existence to the presence of oxygen in the atmosphere and oceans, supplies of which were originally established by a certain type of Prokaryotic bacterium. Indeed, these single-celled life-forms have excreted so much oxygen that surplus supplies have accumulated in the air. The culprits were cyanobacteria – so called because of their blue-green (cyan) colour – which evolved a feeding process now called photosynthesis. They break down carbon dioxide into its chemical constituents, carbon and oxygen, using sunlight, and combine the carbon with hydrogen in water to produce sugars, giving off oxygen as a waste gas in the process. *Prochlorococcus*, a genus of cyanobacteria alive today, is thought to be responsible for as much as *20 per cent of all global oxygen supplies* and is reckoned by some experts to be the most populous creature on Earth with as many as one hundred octillion (10^{29}) individuals alive in the seas.

Stromatolites are rock-like structures that jut up along the shoreline in places like Shark Bay in western Australia. They are sticky to touch because their surfaces are covered in colonies of slimy cyanobacteria that absorb sunlight and give off oxygen. Although they are quite rare today, the world was once peppered with these structures, which first appeared about 2.8 billion years ago, contributing to the massive injection of oxygen into the Earth's atmosphere.

Most microscopic life that thrived in the ancient seas avoided all contact with oxygen as its reactive nature was highly toxic. Prokaryotic life that evolved in an oxygen-free world therefore either had to adapt to an oxygenated environment or find places where levels of oxygen were sufficiently low to ensure their continued survival. Some organisms adapted, learning to feed on oxygen themselves (such as the genus *Rickettsia*). Others linked up, using oxygen as an ingredient for proteins that could glue cells together, to make more complex, multi-cellular creatures that were better adapted for life in oxygenated environments. Others went into hiding – either at the bottom of the sea, hugging close to hydrothermal vents full of sulphur-rich nutrients (such as the methanogens) or, as other lifeforms evolved, deep within the guts of creatures where levels of the gas are low.

Life's original oxygen-hating Prokaryotes live on today inside the oxygen-free innards of animals such as cows and humans, where they feed on ingested food, often providing essential services in return. Colonies of bacteria that live in cows' stomachs break down cellulose, plant material that cows cannot digest themselves. In humans, more than 500 oxygen-hating species of bacteria populate our guts, some of which synthesize valuable chemicals such as vitamin K – essential for blood clotting. Others produce global-warming gases such as methane as a by-product of their own digestive processes, which is why creatures like humans and cows frequently break wind.

The rise of energy-rich oxygen presented awesome opportunities for the evolution of new types of creatures, triggering the genesis of an entirely new domain, the Eukaryotes (see page 36). All forms of higher life – from plants and fungi to fish, birds, mammals and man – are Eukaryotes, whose existence is dependent on adequate supplies of oxygen. Therefore without cyanobacteria, and the photosynthesizing process they originally established, it is hard to see how the creation of a visible, multicellular living world could ever have taken place.

Anthrax

> **FAMILY: BACILLACEAE**
> **SPECIES: BACILLUS ANTHRACIS**
> **RANK: 70**

*A virtually indestructible life-form that will
survive as long as life itself*

IN SEPTEMBER 2001 the United States of America was reeling
from the shocking trauma of the 9/11 terrorist attacks in New York
and Washington in which 2,974 victims died, with another twenty-
four missing, presumed dead. This was a news event like no other
in the history of the modern media, as Bob Stevens, an experienced
sixty-three-year-old photo editor at the *Sun* newspaper, based in
Florida, knew all too well. What Stevens couldn't possibly know was
that, just one week after Islamic terrorists ploughed two fully laden
passenger planes into the World Trade Center and a third into the
Pentagon, an innocent-looking letter would arrive in his newspaper's
mailroom that, for Stevens, was going to prove just as fatal.

All he did was open an envelope. A mixture of brown and white
endospores – looking like mixed-up salt and pepper – spilled on to
his keyboard and puffed into the air. Chances are Stevens thought
nothing of it. Maybe he never even noticed the fine powder during
what was easily the most hectic week in the history of broadcast
news. Ten days later the journalist was in appalling shape. Violent
vomiting and high fever were accompanied by severe disorientation
and complete loss of speech. On Tuesday 2 October Stevens was
admitted to hospital. By Friday he was dead.

Anthrax is a frequently fatal disease caused by the inhalation of endospores from the soil-dwelling bacterium *Bacillus anthracis*. Thanks to modern antibiotics and careful disposal of infected animals that have contracted the disease from spores in the soil, it doesn't kill so many people these days. In 2009 a drum-maker died after inhaling spores carried in imported hides. But only in malicious attacks, such as the yet-to-be-solved biological terrorism of September 2001, are large-scale incidents likely to occur. Eight letters containing spores were sent to US senators and media professionals, killing seven and seriously injuring seventeen. Unlike viral smallpox, this is a disease that cannot easily be passed from one human to another.

Bacillus anthracis matters because it was this strain of bacterium that ultimately led to the discovery by humans that microbes were the source of many deadly infections. In 1875, German medic Robert Koch (1843–1910) published a new theory as to why animals and humans sometimes suffered spontaneous death. After a series of experiments, he was able to reveal the lifestyle of *Bacillus anthracis*. Too much exposure to today's highly oxygenated atmosphere causes this bacterium to become distressed, triggering the process of sporulation by which it shields itself in an almost indestructible microscopic shell that can survive in the ground or air for thousands of years. Koch's experiments showed that ingestion or inhalation of these spores by humans or grazing animals such as cows, horses and sheep cause this bacterium to reawaken in the more favourable, less oxygenated environment inside a host's body. Once in the lungs, the endospores are attacked and engulfed by phagocytes, cells that form a vital part of the body's immune system. But these spores are so resilient that once inside the phagocyte they begin to awaken, germinating into living bacteria that feed off the phagocytes' cytoplasm. Eventually the immune cell explodes, spreading swarms of newly fashioned bacteria throughout the host's body, infecting tissues and causing severe bleeding that rapidly leads to death. Exposed to oxygen in the air through wounds in the host's carcass, sporulation is triggered once again, allowing a vastly increased population of *Bacillus anthracis* to lie dormant in wait, *ad infinitum*, for its next lucky break.

Having proved his germ theory, Koch and his pupils went on to identify different strains of bacteria that were found to be the cause of many lethal diseases including tuberculosis, cholera, diphtheria, typhoid, pneumonia, gonorrhoea, meningitis, leprosy, bubonic plague, tetanus and syphilis. Their research led directly to the development of vaccines and antibiotics that unleashed humanity's own campaign against microbes, dramatically improving human life expectancy. Success in this war has been a major factor behind the exponential rise in human populations from a total of two billion people in the early twentieth century to nearly seven billion individuals today – with all its associated ecological impact.

Being able to bunker down and survive the most extreme conditions through sporulation also makes *Bacillus anthracis* rank highly among organisms most likely to survive in the future, regardless of human or natural catastrophe. These microscopic spores are so difficult to destroy that decontaminating a few US government buildings after the 2001 attacks took more than five years at a total cost of $1 billion. On the Scottish island of Gruinard, used by the British military during World War II to test biological weapons based on anthrax, the clean-up took forty-eight years.

Pseudomonas

> PHYLUM: PROTEOBACTERIA
> SPECIES: PSEUDOMONAS PUTIDA
> RANK: 94

*A bacterium that promises to help solve
the problem of plastic waste*

KAMILO BEACH SHOULD be one of the most picturesque
places on Earth. Reaching out towards the southern tip of tropical
Hawaii's Big Island, this shoreline is a natural mecca for sun-
worshippers. But today no right-minded tourist would willingly go
there because over the last fifty years it has become the most polluted
beach in America, if not the world. The reason, in a word, is plastic.

In the summer of 1997, Captain Charles Moore, heir to a successful Los Angeles oil family, took the short way home after participating
in a prestigious trans-Pacific yacht race. Instead of heading along the
normal sea routes that track the trade winds, he motored his fifteen-
metre catamaran directly through the middle of the Northern Pacific
Gyre, a slowly swirling oceanic circulatory system that rotates clock-
wise in a huge circle from Alaska in the north down the west coast
of the United States and finally across to Asia. What Moore didn't
count on was spending an entire week sailing through thousands of
miles of plastic junk that has been collecting in the middle of this
vortex for more than fifty years.

> *'This part of the ocean is like a giant toilet bowl that never flushes
> . . . Waste that washes up on to Kamilo Beach goes down a foot deep.
> At one time there were toothbrushes, pens, cigarette lighters, plastic*

bottles and caps, but now it's been ground down by the weather into
plastic sand.'

These tiny plastic particles wash into the sea as microscopic fragments, with long-term health risks to both the marine food chain
and other predators, like humans, who feed on fish.

After returning from his yacht race and witnessing the scale of
the impact that modern man is having on the marine environment,
Moore decided to devote his life and wealth to raising awareness
about the problems of plastic pollution. Researchers at the Algalita
Research Foundation, established by Moore, are now exploring
the full extent of 'The Great Pacific Garbage Patch', as the oceanic
vortex of plastic waste off Hawaii is often called. Their latest findings
indicate that the dump covers an area twice the size of Texas, and
contains more than *100 million tonnes* of non-biodegradable plastic.
Most of it has been washed into the sea from landfill sites by the
natural process of erosion and now swirls around in a never-ending
cycle. The problem is getting worse: worldwide plastic production is
increasing so quickly that Algalita researchers say the Garbage Patch
is growing at a runaway rate, expanding tenfold every three years.

Man-made objects are generally designed around the needs of
humans, not nature, and so plastic is durable, not degradable. It is a
remarkable thought that all the plastic ever created by humans still
exists – somewhere. Some of the floating junk in the Pacific Ocean
can even be dated back to the first major phase of plastic production
during World War II. In 2006, a piece of plastic was found in an
albatross's stomach that bore a serial number tracing it to a World
War II seaplane shot down in 1944. According to a United Nations
Environment Programme published in July 2004, more than one
million seabirds are killed by plastic debris each year and, on average, every 2.6 square kilometres of ocean contains 46,000 pieces of
floating plastic junk.

What's to be done? Kevin O'Connor is a microbiologist at the

University of Dublin who thinks he may have an answer to a large part of the problem. For the last four years his team of scientists has been studying the behaviour of a group of Prokaryotes called Proteobacteria and, in particular, a variety called *Pseudomonas putida*. Bacteria are well known for their diverse taste in food. O'Connor believes that with sufficient encouragement, and in the right conditions, this species of bacterium could help solve the issue of plastic pollution by producing a form of biodegradable plastic called polyhydroxyalkonate (PHA for short), which can be used to make everything from plastic wrap and plates to knives, forks and cups. What's more, the food these microbes use to make their own form of organic plastic is styrene, a highly toxic by-product of the manufacture of synthetic rubber.

O'Connor's ultimate goal is to cultivate and, if necessary, genetically modify this bacterium to make the manufacturing of bio-plastics an efficient, economic process. His aim is to turn the fifty-seven million kilograms of styrene waste annually dumped into landfill sites in the United States into thirty-four million kilograms of biodegradable plastic. O'Connor believes his vision could soon make economic sense. In future, companies could sell styrene instead of paying others to dispose of it, reducing the cost of bio-plastic production to about one dollar a kilogram compared with two dollars for a kilogram of artificial plastic today.

The prospect of bacteria reducing mankind's damaging impact on the natural environment is not altogether surprising. Prokaryotic organisms have underpinned the essential processes of all 'higher' forms of life (plants and animals) for hundreds of millions of years. But natural evolutionary adaptations typically take place in geological not human timescales. The dramatic addition of so much artificial waste in the last fifty years – as graphically highlighted by Moore and his researchers at Algalita – has been too quick for most bacteria naturally to adapt. Yet, as O'Connor and his team have demonstrated, species like *Pseudomonas putida* do seem capable of performing the biological equivalent of medieval alchemy. Although their full impact lies in the uncertain future, the potential benefit they offer is awesome: saving the living world from drowning in oceans of non-biodegradable, toxic, man-made plastic junk.

Rhizobia

> **FAMILY: RHIZOBIACEA**
> **SPECIES: RHIZOBIUM LEGUMINOSARUM**
> **RANK: 4**

A bacterium that largely accounts for the successful establishment of human civilizations

BAQUEDANO RAILWAY JUNCTION sits next to a derelict mining town in the scrub of the Atacama Desert in northern Chile. Jutting out from its old engine sheds, which surround a giant disused turntable, are the impressive hulks of several enormous steam trains, perfectly intact. They have been petrifying there, in the scorching sun, since being abandoned more than one hundred years ago. This is like no other place on Earth. There is so little water that humidity actually reaches zero. A few poor locals still manage to get by, squatting in zinc-roofed houses built by British engineers at the end of the nineteenth century when the town was at the centre of a mighty international battle for the control of some of the world's most precious natural resources. Most coveted of all was saltpetre, a natural ingredient for manufacturing explosives and fertilizer vital to European nations such as Germany and Britain during the build-up to World War I (1914–18).

Saltpetre is the by-product of a Prokaryotic process. Over millions of years, symbiotic bacteria from the Rhizobiales order have teamed up with certain plants to convert nitrogen in the air into water-based ammonium and nitrates. These are chemicals that can be absorbed by plants to make proteins, the basic ingredients for cellular life. In return, these plants provide a high-energy diet of carbohydrates for the bacteria, which is essential for the nitrogen-fixing process and their general welfare.

Plants have long since disappeared from the Atacama Desert. The land is so dry that none could grow here now. Yet the bacteria that these long-gone plants once depended on still remain. Like the old steam trains, these precious chemicals – which in most places would have long since dissolved – have been preserved for countless aeons entirely thanks to this region's complete lack of moisture.

Rhizobium leguminosarum is one of the few organisms that is capable of 'fixing' atmospheric nitrogen into soluble nitrates that fertilize the soil so that plants and trees can thrive. It evolved over hundreds of millions of years in association with the legume family of plants, many of them vegetables that have become essential to animal and human diets, such as spinach, beans, peas, lentils, soya and clover.

Roman farmers were the first to note the importance of legumes for nourishing the soil – not that they had any inkling that symbiotic bacteria were literally the root cause of their agricultural prosperity. Luckily, legumes don't just fix nitrogen for themselves. After harvesting, the bacteria in their root nodules leave substantial supplies of excess ammonium and nitrates in the ground, which act as natural fertilizers to support the growth of subsequent crops. Centuries later, Arab agriculturalists, who transported food produce across the vast Islamic empire from China and India to Spain, established the technique of crop rotation in which legumes such as spinach and aubergines were alternated with other crops to ensure the soil was kept rich in nitrates. Medieval European farmers developed their own system of crop rotation using the legume clover to provide essential fertilizer. In the civilizations of Central America, a planting system known as the 'Three Sisters' ensured other legumes, such as beans, were planted alongside gourds and maize, once more ensuring the ground was filled with nitrogen-rich nutrients.

But in the early twentieth century the natural bargain between bacteria, plants and humans began to break down. Two German scientists, Fritz Haber and Carl Bosch, pioneered an industrial process that could artificially 'fix' nitrogen without the need for plants, bacteria or saltpetre. They found that by heating nitrogen and hydrogen gas under high pressure they could synthesize large quantities of ammonium, so that by 1915 a new age of artificial fertilizers was born. Now any country could make its own supplies of gunpowder and fertilizers using nothing more than fossil fuels to generate the heat required to run the process. Imperial dependency on saltpetre from the driest deserts of the world (India and Chile were the main sources) was abandoned – along with the steam trains of Baquedano that once transported huge volumes of the stuff from inland mines to ports on the coast.

The historic impact of nitrogen-fixing *Rhizobium leguminosarum* and its several closely related Prokaryotic varieties is beyond doubt. Without them, the Neolithic revolution from which human agriculture has its origins (see On Agriculture, page 201) would most probably have stalled through soil exhaustion, throwing into doubt the long-term viability of settled human civilizations. Its other big impact began in the mid-ninth century when Taoist monks in China discovered how to manufacture gunpowder from naturally occurring nitrates (saltpetre). But future prospects for a revival in widespread organic nourishment for the world's soil now look brighter. Scientists and governments alike are aware of the environmental side-effects of modern food production systems on the Earth's climate and the eventual depletion of fossil fuel supplies. The Haber process alone uses as much as *5 per cent of the world's total production of natural gas*. How much more safe, efficient and cost effective are nature's bacterial systems that use none! Artificial fertilizers also acidify the soil and release additional greenhouse gases such as ammonia and nitrogen dioxide (NO_2) into the atmosphere.

Traditional organic methods that rely on nitrogen-fixing bacteria and their legume hosts are the only known long-term source of soil fertilizer. All organic farming depends on them. Therefore the story of *Rhizobia*, the nitrogen-fixing bacterium, is likely to have a future just as important as its past – even if, for a few eccentric years, it went briefly out of fashion when the world went fossil-fuel mad.

3

On Symbiosis

*How genes and cells converged,
establishing a new set of evolutionary
rules that led to multicellular life.*

Slime Mould

PHYLUM: MYCETOZOA
SPECIES: DICTYOSTELIUM DISCOIDEUM
RANK: 74

*The surprising origins of intelligent,
self-organizing systems*

IT MUST HAVE BEEN one of the most bizarre announcements in the history of scientific research. In the autumn of the year 2000, Japanese scientist Toshiyuki Nakagaki told the world how a slime mould could successfully navigate its way through a maze by the shortest possible route. It was, he claimed, a demonstration of the most primitive form of Earthly intelligence. Working with a team of researchers at the Biomimetic Control Research Centre in Nagoya, Nakagaki dotted several pieces of slime throughout a three-centimetre by three-centimetre maze constructed on top of a plate of agar. Over the course of a few hours the blobs grew tube-like structures called pseudopodia that caused the whole maze to fill with a long, sticky column of slime. There were four possible routes through the maze. Food was placed at both exits. After eight hours the column of slime had shrunk so that its 'body' filled only the parts of the maze that formed the shortest route from one piece of food to the other. Remarkably, it had worked out how to organize itself into the most efficient possible shape for foraging for food and in the process solved the puzzle of finding the quickest and most direct passage through the maze.

Slime moulds are living descendants of some of the earliest forms of Eukaryotic life. Since their modern discovery in a North Carolina forest in 1935, scientists all over the world have been fascinated by their extraordinary behaviour. Slimes are so puzzling that they have variously been classified as plants, animals and fungi. The consensus today is that while these creatures share some characteristics with all three, they rightfully belong in a separate kingdom altogether, called Amoebozoa. Recent genetic sequencing of the slime mould *Dictyostelium discoideum* confirms their evolutionary significance. They split off from the Eukaryotic tree very early on, just after the initial division between plants and animals and before the first sign of fungi.

The peculiar life cycle of the slime mould brilliantly demonstrates how versatile life became once symbiotic cells (Eukaryotes) had evolved new rules for collaborative living. As the slime mould maze experiment shows, the emergence of what humans understand as basic intelligence starts right here. The slime life cycle begins when spores hatch out on moist ground, growing into single-celled Eukaryotes that graze on bacteria in the soil or on rotten logs. Their flexible cell structure allows them to surround their prey, engulfing and then digesting it as food. This process, called phagocytosis, is of special interest to modern science as it represents the earliest appearance in nature of tactics later employed by vertebrate animals to fight disease (see shark, page 71). White blood cells in our bodies behave just like individual slime cells, engulfing and digesting bacterial invaders to neutralize any threat they pose to the body. Recent genetic sequencing of *Dictyostelium discoideum* shows that the genes responsible for attacking disease in humans are essentially the same as those found in this much more ancient slime mould.

Slime's real eccentricity becomes apparent when food runs short. Individual cells secrete a chemical (adenylyl cyclase) that acts like an alarm informing other nearby slime cells that since food is scarce, genetic survival depends on collaboration. They do not engage in a war over scarce resources. Instead, they group together in an orderly manner to form a completely new multicellular creature – a type of slug – from scratch. For years scientists thought there must be some master cell co-ordinating the process – if only they could observe this bizarre behaviour closely enough. But it is now clear that slime cells represent a natural self-organizing system. The collective actions of many individuals following a simple set of collaborative rules result in intelligent group behaviour.

As soon as a cell emits the chemical alarm, its neighbours either move towards the chemical or amplify its effect by secreting more alarm signals. In this way, moving in highly regulated six-minute pulses, up to 100,000 cells can rapidly swarm to form a single body of multicellular slime. The aggregate organism substantially increases the chances of successfully finding food, as ably shown by the slime in the Japanese maze, since as a collective its size, reach and powers of movement and detection are so much greater than those of any one individual. When conditions are just right, the slime mould begins to metamorphose once again. Within about ten hours it has turned itself into a fruit. Some cells then sacrifice themselves to form a dead stalk made of cellulose that raises up other cells at the front that have turned into balls of hardened spores ready to be blown away by the wind. Like the bacterium anthrax (see page 23) these well-protected cells can survive for years, waiting for a favourable moment to begin germination, starting the slime's life cycle all over again.

The significance of slime today lies in the insight this extraordinary life-form has given into the workings of evolution and emergent self-organizing systems. Intensive study of *Dictyostelium discoideum* since the 1960s and the sequencing of its genome (completed in 2005) has yielded a fresh understanding of how complex intelligent systems arise without the need for centralized direction. In the last fifty years, humans have begun to mimic these systems and a new branch of science called 'artificial life' has emerged as a direct result. Open-source software development, evolutionary computer games, the self-organizing maintenance of complex communications networks – even the rise of the Internet – are just a few examples of the recent human application of natural systems that first evolved in the ancestors of today's slimy moulds, hundreds of millions of years ago.

Water Mould

CLASS: OOMYCETES
SPECIES: PHYTOPHTHORA INFESTANS
RANK: 30

How a microscopic protoctist reshaped history

PROTOCTISTA, WHICH INCLUDES such diverse groups as amoeba, algae and slime, is one of the most ancient and least well known divisions of Eukaryotic life. Yet the influence of some of its species during recent human history (especially for those of Irish descent) has been especially profound. Of particular note are the Oomycetes, microscopic unicellular creatures, more commonly known as water moulds.

Talented French botanist Heinrich Anton de Bary first described the parasitic creature *Phytophthora infestans* in 1861. Among his many weird fascinations was mixing tiny spores, removed from the leaves of potato plants, into drops of water, which he placed under a microscope. He was astonished to see that the spores germinated and multiplied, each growing their own tails and swimming about as if they were miniature tadpoles, just hatched out from eggs (see plate 2, top). The more he watched, the more de Bary realized these were no normal vegetative fungi. When he saw how these miniature animals behaved on the surfaces of potato leaves he was amazed. Tubes seemed to grow out of the organisms, punching their way through the leaf's surface. A creature would then squeeze its way through the opening, leaving an empty, dead layer of skin behind. Once inside, it would swell, elongating and branching its way from cell to cell, helping itself to the plant's nutrients until the leaves were completely consumed. Eventually, the mould would fruit by thrusting its roots back up through the pores of the stricken leaf, so that its spores could be borne by the wind towards another nearby plant. De Bary quickly realized the significance of what he had observed – which is why in 1861 he named these microscopic creatures *Phytophthora*, meaning 'plant-eater' in Greek. Just sixteen years earlier, the most devastating famine in modern times had struck the Western world. De Bary had at last worked out its cause.

In mid-nineteenth-century Europe potatoes were still not accepted as respectable food on the dinner tables of many countries, associated as they were with the 'savage' natives of Central and South America (see potato, page 210). But this was not the case in Ireland. Laws passed by the ruling English parliament prohibited natives from owning more than a few acres of land each, so high-yielding potatoes had become a key part of poor people's diets. Tubers from these plants provided everything someone with only a little land could wish for. They were simple to grow, easily stored and full of proteins and vitamins B and C; a diet of potato mash with milk proved sufficient to feed a large family, and all its livestock, throughout the year. It was largely thanks to the potato that, within 100 years of its introduction, the population of Ireland rocketed from three to eight million.

Meanwhile, more than 6,000 miles away in the mountains of Peru, the little protoctist *Phytophthora infestans* was enjoying its own population boom, this time feasting on potato plants grown by the indigenous agriculturalists of South America. Wisely, the ancestors of these industrious farmers cultivated hundreds of different potato varieties. This meant that if any one type succumbed to disease, other more resistant varieties could still be successfully farmed in their place. Echoing the natural benefits of sexual reproduction, their pursuit of biodiversity provided an insurance policy against devastating infection. In the early 1600s, when Spanish conquistadors subdued the Incas of South America, it seemed necessary to export only a few popular types of vegetables as curiosities back to Europe. Little did these opportunists realize just how valuable these weird-looking, knobbly tubers would eventually become. Nor did they appreciate the consequences of exporting only a handful of varieties, such as 'Lima', 'Peruviennes' and 'Cordilieres'.

It wasn't until the early 1840s that diseased tubers crossed the Atlantic Ocean in ships from North America to Europe, triggering outbreaks of late blight in Belgium, Germany and Holland, beginning in 1845. No country suffered quite like Ireland, so dependent were its people on their simple diet of potatoes and milk. After the failure of three harvests (1845, 1847 and 1848), the country was devastated. More than one million of its people are thought to have died directly of starvation and disease brought on by malnutrition. Desperate to escape the appalling conditions of the worst famine to affect Europe since the time of the Black Death of 1348, millions of Irish fled, driven off the land by laws that made anyone who owned more than a quarter of an acre ineligible for aid. Most went to America, although some settled in Canada and Australia. The impact on the Western world has been immense. The United States now has thirty-six million Irish Americans, 12 per cent of its entire population. Not that *Phytophthora infestans* was the only cause or indeed the initial trigger of Irish emigration to the Americas, which began soon after the Napoleonic Wars (1804–15), yet its arrival in 1845 massively increased the flow.

The effects of blight and its associated human emigrations have been felt ever since, from the global spread of Catholicism (two thirds of Irish immigrants were Catholics) to an enduring bitterness between many English and Irish people over the failure of the British government to help its own people, a crime that some have since called genocide. The blight triggered demands for Irish independence, leading to the Irish War of Independence (1919–21), the partitioning of the country into north and south and the founding of an independent Ireland, ratified by the Anglo-Irish Treaty of 1921, a status quo that persists to this day.

On Symbiosis

Late blight in New Zealand in 1905 was one of the last water mould epidemics, thanks to the effectiveness of a pesticide invented in 1885 called Bordeaux Mixture, although Chinese potato plants remained unprotected as late as the 1950s. With the birth of systemic pesticides following World War II, it seemed the story of potato blight – *Phytophthora infestans* – had finally come to a close. But never underestimate the power of Eukaryotic sexual reproduction to secure the survival of a species. In the 1980s, a resurgence of new varieties of *Phytophthora infestans* gave the farming communities of America and then Europe a nasty shock. Thanks to sex, new strains of water mould (they are capable of both sexual and asexual reproduction) began to appear with sufficient genetic diversity to benefit from at least some natural immunity to even the most persistent onslaught from artificial pesticides. In 1984 a sexual variant (called A2) that originated in Colombia was spotted in Switzerland and has since been adventuring around Europe. To the utter dismay of the potato farming community, it reached Ireland in 1989. During its progress, *Phytophthora infestans* has been mating with other strains, producing genetically unique specimens of enormous diversity, some of them offering resistance to traditional cures for late blight disease.

So blight is back. *Phytophthora infestans* is once again being described as 'the world's worst agricultural disease', costing farmers billions of dollars a year to contain. A new fightback is now under way by genetic scientists splicing genes taken from a wild South American potato plant *Solanum bulbocastanum* that has, over an unknown number of generations, developed its own natural immunity to these persistent protoctists. Genetically modified, protoctist-resistant potatoes may eventually become a reality, if only the world's seed stock can be comprehensively updated. But, in true Eukaryotic tradition, how long will it be before another sexual deviant saves its species from destruction by humans?

Algae

FAMILY: DICTYOSPHAERIACEA
SPECIES: BOTRYOCOCCUS BRAUNII
RANK: 2

How scum could power the planes of the future

DAVID DAGGETT IS a man with a mission. His goal is no less than to save the worldwide aviation industry from collapse. Volatile fuel prices and global concern about carbon emissions are causing people and governments alike to think twice before they fly. As Technology Leader for Boeing Aerospace's energy and emissions product development group, Daggett's task is to identify a new long-term source of aviation fuel that's plentiful, cheap to produce and kind on the environment. Thanks to his own passion for growing microscopic algae at home, Daggett believes he may just have found the solution.

Algae are among the oldest and most diverse organisms on Earth – like slime and moulds. They are Eukaryotes and are classified today in the kingdom of Protoctista. Early-nineteenth-century scientists divided algae into groups based on their different colours. Green algae are the ancestors of all land plants and trees. Red algae range from large seaweeds to microscopic varieties that attach themselves to the skeletons of coral fish, turning their reefs pink (see page 52). Brown algae, the third major group, form vast marine forests of seaweed called kelp that grow as fast as half a metre a day. Algae are prodigious producers of oxygen, having gained a capacity for photosynthesis from an ancient symbiosis with cyanobacteria (see page 20). Without marine-based algae today global oxygen supplies would quickly dwindle. Collectively they produce a far greater quantity than all the land plants and trees combined.

Without the countless varieties of microscopic algae, larger forms of sea life would never have been able to evolve, let alone survive today. Small filter-feeding marine animals such as krill, clams, mussels and oysters depend on hundreds of thousands of different species of microscopic algae that float through the water, as do larger creatures such as flamingos, basking sharks and even the enormous baleen whale. Plankton, the collective name for many of these algae, lies at the base of the entire aquatic food chain. Land life would have been just as impoverished without marine algae. Ancestors of the lungfish, creatures that masterminded the escape from the seas, would have found precious little to feed on in their newfangled terrestrial habitat without land plants – all of which are descended from ancestral green algae from the seas (see page 44).

Like many of the earliest Eukaryote life-forms, algae are highly collaborative creatures. Lichens, corals (see page 52) and sponges (see page 48) have survived for hundreds of millions of years chiefly thanks to the services of photosynthesizing algae that provide them with plentiful supplies of oxygen and sugars.

Algae have played a key role in human history too. Various types provided ancient human civilizations with a source of highly nutritious food. Edible kelp called Kombu (*Laminaria japonica*) has been eaten for thousands of years by cultures in the Far East – today it is still a vital source of food in countries such as Japan, where it is used to add flavour and nutritional value to soups and broths. Other varieties of kelp have been used in the relatively recent past to make potash, a necessary ingredient in soap and glass manufacture. In other places algae has become an important fertilizer. Maerl is a type of red algae found on the seabed off the coast of Cornwall, now used as an organic soil fertilizer in place of artificial nitrates.

Other types of microscopic red algae called coccolithophores and foraminifera form hard outer shells by absorbing carbon dioxide from the atmosphere and turning it into calcium carbonate (see page 52). As part of this process they have a direct influence on the Earth's all-important water cycle, a process which has made land life and the evolution of humanity possible. A waste gas produced by these algae, called dimethyl sulphide, helps seed clouds by providing perfectly sized surfaces around which water vapour can condense back into liquid droplets. Apart from delivering fresh water supplies around the world, clouds also play an essential role in helping to cool the Earth, making it suitable for life. Much of the Sun's infrared radiation reflects off the tops of clouds back into space, so helping to regulate global temperatures. Indeed, the impact of this diverse family of creatures on human history goes further. When these tiny organisms die, they sink to the ocean floor and, over millions of years, get crushed into a type of sedimentary rock called limestone, which has been one of the most popular building materials used by humans ever since the rise of the Egyptian pyramids more than 5,000 years ago.

It is the future bounty of particular types of algae that experts including Dave Daggett think warrant closest attention. Some, including *Botryococcus*, have the potential to help save the modern world from a global warming catastrophe by weaning us off our dependency on fossil fuels such as coal, oil and gas. 'Algaculture', the practice of farming algae for oil production, has ancient natural origins. Kerogen is a mixture of organic compounds buried deep underground that is made up of decomposed marine algae, cyanobacteria and other protoctists. When it is heated to the right temperature in the Earth's crusts it releases crude oil or natural gas. Just imagine being able to accelerate that same process which normally takes hundreds of millions of years into real-time production – that's the potential of algaculture.

As a result, governments, energy companies and opportunistic entrepreneurs are now showing a great deal of interest in the potential of this sustainable and environmentally friendly energy source.

On Symbiosis

Something of a green gold-rush is unfolding with speculators large and small flooding into the market, chasing the prospect of potentially limitless biofuels without the drawbacks seen with crops that require large amounts of land, water and artificial fertilizer. Some companies are even experimenting to see whether pumping the carbon dioxide emissions from power-station smoke stacks directly into tanks of algae can convert this greenhouse gas into materials that can be harvested as bio-diesel. Others are turning to genetic engineering, tinkering with the creation and patenting of algae species that are faster reproducing, higher yielding and easier to harvest in order to help make the mass-production of bio-diesel more profitable.

Another possibility is to use algae as an environmentally friendly way of producing hydrogen – a potential fuel source for planes, trains and automobiles. When certain species of algae, such as *Chlamydomonas reinhardtii*, are deprived of sulphur they switch from producing oxygen as a waste product to producing hydrogen, and in such volumes that, according to one recent scientific paper, it could make the process commercially viable. In other trials, species such as *Sostera marina* are being harnessed by Japanese scientists in an experiment to construct giant carbon sinks that can absorb vast quantities of carbon dioxide, turning this problematic climate-changing gas into organic fertilizers and fuel. In a newspaper interview project leader Dr Masahiro Notoya from the University of Tokyo put the case for the impact of algae, past, present and future, rather succinctly:

> '*It's actually thanks to seaweed that we're here at all. When the world was young, it was the little blue-green algae and other seaweeds that, over the years, converted so much of the carbon dioxide in the air into oxygen and eventually pushed it up to the levels it is at today. Now that the balance is being thrown off, it's time for the seaweed to come and help again.*'

Sponge

> PHYLUM: PORIFERA
> SPECIES: MICROCIONA PROLIFERA
> RANK: 72

*A primitive animal survivor that pioneered the
art of multicellular construction*

MOST PEOPLE NATURALLY associate a sponge with having a bath. Not so Henry Wilson, an early-twentieth-century professor of biology at the University of North Carolina, who instead used a sponge to conduct one of the most illuminating and bizarre biological experiments of all time. Sponges are classified as primitive animals. Rooted to the sea floor, they use tiny internal hairs to beat sufficient water through their mostly hollow bodies to extract the microscopic algae and plankton on which they live (see plate 2, bottom). Simple and motionless they may be, but to Wilson sponges were his complete *raison d'être*. He spent his whole life studying them. Modern science owes a huge debt to Wilson's obsession. To the great surprise of his biologist colleagues, in the early summer of 1907 Wilson demonstrated conclusively that the apparently simple sponge can perform a rather remarkable conjuring trick. His demonstration went as follows:

First, Wilson minced up a silicate sponge (*Microciona prolifera*) by ramming it through a sieve until all that was left was a pile of individual, crumb-like cells. Next, he placed the sponge cells in a dish of water and stained them red so that he could see what, if anything, would happen. Then he watched.

To begin with, the cells just seemed to jiggle about independently at random. Gradually he saw them begin to gather together, forming a red cellular clump. Then the cells began to take on various shapes and patterns. Finally, the cells reorganized themselves into an exact replica of the sponge that Wilson had previously pushed through the sieve.

What Wilson witnessed was the extraordinary process by which nature builds multicellular bodies. The natural emergence of self-organizing biological systems is the process by which an entire organism is built out of independently operating cells from scratch – and all without the need for any form of central command. Some people say this is the basis of all 'intelligent' forms of life. After Wilson's experiment, scientists have never been so dismissive about sponges again. Indeed, this classic sponge experiment has been instrumental in the development of a specialist branch of body-building biology called embryology, which now provides some of the most important insights into the overall foundations of evolutionary science.

In-depth analysis of the sponge experiment shows that just *three* basic processes lie at the heart of what makes multicellular life possible. First is the need for some kind of glue that helps individual cells bind together. Plant and animal cells secrete special proteins that do this job. Collagen is one such, also found extensively in humans. It helps to keep the *two trillion individual cells* in our body stuck firmly together. Secondly, cells have different types of shapes and surfaces by which they attach themselves to each other. These rivet-like structures are key to making sure individual cells organize themselves correctly, latching on to other cells in the right order. Finally, given that there is no overall conductor or chief in charge of the operation, it is essential that each cell can communicate with its neighbours, passing on instructions such as whether to (a) die, (b) divide, or (c) issue further instructions to other nearby cells. Again, they secrete special proteins to carry out these necessary communication functions.

What on Earth Evolved? . . . in Brief

A sponge may seem an absurdly simple creature compared to a reptile, bird or human, but in fact the cellular routines that it uses to build and (if sieved up) rebuild itself are similar to those used by a human embryo in its mother's womb. All multicellular organisms (which includes all plants and animals) use the same basic self-organizing system as a sponge, allowing their cells to collaborate in a way that maximizes the creature's overall chances of survival and reproductive success.

While sponge-like creatures have been detected in the fossil record as far back as 600 million years ago (about fifty million years before creatures in the sea developed bones, shells and teeth), there still remains the question of why it took so long for multicellular life to emerge. After all, the first traces of Eukaryotic life appear about *one billion years* before the first seaweed and sponge-like fossils.

The reason for this delay is now thought to be down to levels of atmospheric oxygen. Collagen and other proteins that provide the glue that help cells stick together can be manufactured only in oxygen-rich environments. Therefore only when levels of oxygen increased sufficiently (due to photosynthesizing cyanobacteria, see page 20) could multicellular bodies begin to develop. By 600 million years ago visible life-forms were finally filling up the oceans – seaweeds, jellyfish, sponges and worms being among the first – leaving their shapes for posterity as occasional fossils in the rocks. Once creatures like sponges adopted the habit of linking up their cells into giant collectives – of growing big – it is safe to say that life on Earth was never the same again.

4

On Sea Life

*How biological variety and
constantly changing environments
caused an explosion of new species
to evolve in the seas.*

Stony Coral

ORDER: SCLERACTINIA
SPECIES: POCILLOPORA DAMICORNIS
RANK: 7

One of nature's oldest and most colourful species,
founded on a mutual collaboration

'MULTIPLY, VARY, LET *the strongest live and the weakest die.'* So said Charles Darwin as he boldly tried to wrap up his life's work by boiling the behaviour of the natural world down to a single general law.

Observers wishing to challenge this claim would be wise to consider the case of corals – organisms that evolved near the start of the saga of multicellular life about 600 million years ago. Cnidaria, the phylum to which jellyfish, anemones and corals belong, shows up clearly in the earliest animal fossils.

One leading naturalist recently declared that corals are ecologically speaking '*surely among the most significant of all animals*' chiefly owing to their capacity to build underwater mountains. More than one million square kilometres of coral reefs are thought to exist in the world's oceans. Each reef, standing in waters up to fifty metres deep, is mostly composed of the dead skeletons of tiny organisms called polyps, related to jellyfish, which secrete hard calcium carbonate coatings around their soft bodies as a simple form of protection.

Although these reef-building coral polyps can feed using tiny stinging tentacles (like their relatives the jellyfish), most of the time their energy is supplied through a symbiotic collaboration with a genus of microscopic red algae called *Zooxanthella*. More than 90 per cent of the polyp's energy requirements come from the ingestion of these algae that then reside inside the polyp's body tissue turning sunlight into energy via photosynthesis. In return, the polyp protects the algae and supplies the carbon dioxide it needs to photosynthesize.

When polyps die, the deposition of their calcium skeletons, layer upon layer, creates huge reefs, and eventually enormous underwater mountains, providing an essential habitat for both living polyps and algae. This ensures that the algae remain as close as possible to the sea's surface, where the light from the Sun can easily penetrate. The red algae also secrete calcium carbonate in layers, adding their own contributions to the structure of a successful, thriving coral reef. Such close interdependent co-operation between species is known as mutualism.

The Great Barrier Reef complex off the coast of north-east Australia is easily the world's most massive structure made by living organisms. More than 2,900 individual reefs and 900 islands stretch across 2,600 kilometres of open, shallow seas, covering an area of approximately 350,000 square kilometres – that's nearly twice the size of Great Britain . . .

Scale and size aside, the impact of the symbiotic union between red algae and these tiny jellyfish-like animals that first appeared in the Cambrian Period (542–488 million years ago) has been immense. Repeated reef-making eras throughout geological history have harboured the richest diversity of marine species. Over 4,000 different species of fish inhabit today's healthy reefs, making their homes in cracks and crevices that provide perfect breeding grounds for all manner of marine creatures. It's not just fish that benefit. Reefs also provide excellent homes for sponges, jellyfish, worms, crustaceans, molluscs, starfish, sea squirts, sea turtles and sea snakes.

Until recently, mammals rarely paid them a visit, except, perhaps, for a few curious but friendly dolphins. While aboriginal Australians have regularly fished from the reefs off Queensland for the last 40,000 years, the modern threat to coral reefs comes from rising levels of carbon dioxide in the air, which dissolves in the sea making it more acidic. Polyps eject symbiotic algae from their systems in response to sea temperature rises and ocean acidification, a process that leads to coral bleaching. Feeding carnivorously using their stinging tentacles is an alternative but temporary survival strategy. Unless conditions revert, the reefs eventually die.

As many as 10 per cent of the world's reefs are thought to have perished in the last hundred years because of the impact of human-induced climate change. Further damage to these jewel-like marine ecosystems has come from the introduction of modern artificial fertilizers that run off farmland into the seas, increasing nutrient levels in the water, causing the wrong type of algae to grow. This has the effect of blocking out sunlight from the water's surface, smothering a reef to death. Tourism is another hazard. Since the end of World War II, coral islands, with their attractive climates, beautiful turquoise-blue seas and exotic marine life, have become some of the world's most popular holiday resorts (for example, in the Maldives, along the Red Sea, Indonesian and Mexican/Belize coasts).

The process of evolutionary mutualism that gave birth to life's extraordinary coral reefs does not mean Darwin was wrong in his claim that the modification of species (evolution) occurs through a perpetual competitive struggle for survival. It's just that, as the symbiosis of coral fish and red algae shows, in certain environments relationships between different species are just as likely to be founded on mutual benefit rather than competitive pressure (see also oak, on page 142, for an example of mutualism on the land). Coral reefs are powerful places for the natural conservation and co-operation of species, which is largely why the lifestyles of polyps and their symbiotic *Zooxanthella* partners have survived so well to this day.

Roundworm

PHYLUM: NEMATODA
SPECIES: CAENORHABDITIS ELEGANS
RANK: 20

*A super-terrestrial life-form that stretched
survival to its limits*

WHEN SPACE SHUTTLE *Columbia* broke up on 1 February 2003 some sixty-one kilometres above the Earth's surface, no one expected search teams to find any signs of life in the wreckage. Amazing as it sounds, despite the inevitable and tragic loss of human life, we now know that they did.

A team of scientists headed by Dr Catherine Conley from a NASA research centre in California successfully recovered five canisters of *Caenorhabditis elegans*, a species of roundworm that had been taken on board the mission (ST-107) as part of an experiment designed to discover more about the effects of zero gravity on organisms living in space. After painstakingly sifting through the wreckage for seven weeks, five out of seven canisters were recovered. In all but one container the worms had survived their highly traumatic re-entry into the Earth's atmosphere as well as a colossal and mostly unprotected impact with the ground, crashing into the surface at speeds estimated at more than 600 miles per hour.

This extraordinary story underlines the incredible resilience of certain small animals, such as these nearly microscopic worms. More than 80,000 different species of roundworm are currently known to science. Most feed on microscopic algae and fungi or dead organic matter, decomposing and recycling nutrients in the soil and seabed.

A few (about 15,000 species) live as parasites in other hosts, some-times causing diseases in humans, farm animals, cats, dogs and plants. Roundworms are among the most numerous and least fussy creatures on Earth, adapting to life in almost every conceivable habi-tat from the Arctic to the tropics and from the tops of mountains to the deepest mid-ocean trenches. The number of undescribed species is thought to run into millions. Creatures belonging to this phylum represent a staggering *90 per cent of all life found on the sea floor*. On land, literally billions are contained in just one acre of topsoil, and as many as 90,000 roundworms are said to be able to occupy a single rotting apple.

Today's roundworm ancestors were among the first creatures to establish a fully developed three-layer (triploblastic) system of body design. Unlike jellyfish and corals (cnidarian), which have only two layers of cells (diploblastic), more sophisticated creatures later exploited the three-layer design to support the emergence of internal organs, circulation, muscular activity and a skeleton with-out disrupting the all-important and ongoing processes of internal digestion.

As well as being able to survive a space crash, *Caenorhabditis elegans* has another special significance. It was chosen by Nobel-Prize-winning biologist Sydney Brenner as a model organism for detailed scientific research into how animal bodies work and how their cells specialize to take on different tasks and functions (round-worms have a primitive nervous system as well as a digestive tract). He chose well. *Caenorhabditis elegans* has proved ideal for intensive scientific study, being small, simple and easy to cultivate. Its CV also includes coming fully back to life after being frozen indefinitely in liquid nitrogen. Once thawed, on it goes, feeding, excreting, breeding more of its kind. So venerated has this species become in research fields that it was awarded the honour of becoming the first creature whose genome was fully sequenced. Scientists published their results in 1988. The subsequent publication of the Human

Genome in 2003 showed rather surprisingly that humans and roundworms share approximately the same number of genes (about 20,000–25,000).

The biological resilience and survival capacity of *Caenorhabditis elegans* following its remarkable journey into space and back has recently boosted theories for the extraterrestrial origins of life – a branch of science called astrobiology. If small animals such as roundworms can successfully travel through space and land on a planet at more than 600 miles per hour, then who knows what could have arrived on the Earth on meteorites hundreds of millions of years ago, seeding life or at least shuffling its course into new, unexpected directions?

Trilobite

PHYLUM: ARTHROPODA
SPECIES: OLENELLUS SP.
RANK: 31

A family that switched on the lights of life

CHARLES DARWIN WAS always anxious to remind his readers of the old dictum that 'nature never moves in leaps' (*natura non facit saltum*) – indeed his whole theory of natural selection was built on the idea that gradual, incremental changes from one generation to another are what have caused, from so simple a beginning, '*endless forms most beautiful to have been evolved*'.

It is still with some courage that modern scientists challenge Darwinian orthodoxy. One American naturalist, Stephen Jay Gould, famously tried to destroy the idea of incremental evolutionary change in a detailed study of the extraordinary fossils from the Burgess Shale in Canada. His book *Wonderful Life*, published in 1989, proposed that chance rather than fitness was the major agent of evolutionary change and that exotic creatures etched into the Burgess Shale were often more various and therefore more 'evolved' than those that have followed. A big debate still rages today between scientists as to whether evolution is a gradual or a more punctuated process. Australian biologist Andrew Parker has recently made a powerful case for at least one profoundly punctuated moment in the evolution of animals which, if correct, puts the impact of one type of extinct creature right at the top of nature's most significant species.

Trilobites were marine arthropods (invertebrate creatures with jointed legs) that looked rather like large, heavily armoured woodlice.

On Sea Life

Although they have been extinct for more than 250 million years, their arrival on to the stage of life at the beginning of the Cambrian Period 542 million years ago appears to have caused a major evolutionary upset. Before trilobites, most animals were worm-like, although a few, like coral polyps, had started to experiment with sponge and jellyfish-like forms. At about the time the last 'Snowball Earth' ice-age periods retreated (575 million years ago), sea creatures began to grow larger probably because of rising oxygen levels in the atmosphere. According to Parker the increased intensity of sunlight that penetrated through the melting ice became a significant selective pressure in favour of creatures that could determine light from shade. Swimming towards the light meant floating upwards nearer the surface of the sea where photosynthesizing food sources, such as algae (see page 44) were most abundant. Mobile creatures sensitive to the light therefore thrived.

Parker believes this increasing luminescence was responsible for the birth of vision. *'With our eyes open suddenly we see the world very differently . . . The light switch was turned on for the first and only time – and it has been on ever since.'* Once one type of organism learned how to see, through myriad calcite lenses connected to a series of internal nerves, evolutionary pressures in the sea were affected all around. Trilobites quickly became supreme rulers of the marine world. Fossil evidence makes it hard to tell exactly which of the 6,000 trilobite species so far identified by palaeontologists were life's first visual pioneers, but one reasonable candidate is Olenellus, a common and highly successful early Cambrian trilobite. Its fossils show that it definitely had eyes.

Within just a few million years, almost instantaneously in geological time, creatures began to evolve elaborate defences against these new visionaries of the sea. By *c.* 530 million years ago, a plethora of creatures from a number of different animal phyla started to appear in the fossil record as if they had arrived from nowhere – the famous so-called 'Cambrian Explosion'. These were creatures that had been gradually evolving for millions of years, but now survival meant having effective defences against all-seeing trilobites.

The construction of hard outer shells (which happen to fossilize easily) was one strategy pioneered by brachiopods (lamp-shells) and soon followed by the bi-valvular molluscs (ancestors of today's clams, mussels, oysters and scallops). Other strategies included establishing defensive barriers. Fossils of many creatures from the Cambrian era feature such defences, including the velvet worm Hallucigenia (see page 62), and bristle worms Canadia and Wiwaxia, oval-shaped creatures covered by overlapping shields and long, outward-projecting swords.

Other worm lineages evolved instinctively to bury themselves into the seabed or between cracks in the rocks. Some became so well camouflaged (such as the orange-coloured velvet worm called Asheaia) that even those equipped with eyes could not distinguish them from sponges rooted to the floor. Successful species, once similar in appearance, were those whose offspring had an advantageous shape, colour and form. An explosion of variety reached its peak.

The supremacy of trilobites did not last for ever. They too developed defences, initially to protect against being attacked by each other (hence their own evolution of hard outer skeletons and body armour) but later to protect them from other arthropods that had developed vision, chiefly the fearsome Anomalocaris, with its two bulging eyes and menacing pair of grasping tentacles. Growing as large as one metre long, this creature became one of the most powerful carnivores in the later Cambrian seas.

During the next evolutionary period (the Ordovician), creatures across other animal groups independently began to evolve visual senses, including annelids (bristle worms), cnidarians (box jellyfish), molluscs (such as the squid and octopus) and chordates, the phylum that eventually evolved into all vertebrate animals including sharks, fish, amphibians, reptiles, birds and mammals. So powerful a technology was the trick of sight that today over 95 per cent of all animal species have eyes.

The need to survive in a world where some had vision led to the development of countless new species with dramatically different outward forms and instinctive behaviours. It also led to the construction of hard outer bodies, giving humans an invaluable fossil record with which to puzzle over the ancient past.

Velvet Worm

PHYLUM: ONYCHOPHORA
SPECIES: PERIPATUS SP.
RANK: 56

Baby steps that became prehistory's biggest leaps

NEIL ARMSTRONG MADE the most famous ever 'leap' for mankind when, on 21 July 1969, he stepped out of America's Apollo 11 lunar capsule to become the first man to walk on the surface of the Moon. An even greater achievement, however, was made by the much smaller steps of another creature which, in a bid to escape the pressures of living in the increasingly hostile seas, tiptoed ashore about 480 million years ago.

Life in the seas at the beginning of the Cambrian Period 542 million years ago became a great deal more brutal. Creatures that thrived were those that successfully adapted to fend off far-sighted predators such as the war-like trilobites (see page 55) or the metre-long carnivorous menace Anomalocaris. Some of these adaptations appeared so strange to scientists that for a while it seemed as if whole new families of now extinct creatures suddenly had evolved out of nowhere. Now, however, regardless of their bizarre looks (see plate 3, top), these creatures are thought to be, as Darwin predicted, ancient common ancestors of animal families that still populate the world today.

Among the most bizarre-looking, and for a while most misunderstood, life-form was the marine caterpillar-like creature called Hallucigenia. On the underside of its long, worm-like segmented body were what were thought to be a series of long, stiff legs, while on its topside was a brush of soft upward-pointing tentacles that groped in the water for food. Nothing like this weird creature seems to have existed before or since. That, at least, was what everyone thought until Swedish dentist-turned-palaeontologist Lars Ramskold discovered that experts had been looking at its fossils upside down! When turned the right way up the long, stiff spines could be seen to be layers of heavily armoured protection against attack; meanwhile the tentacles were instantly transformed into soft, stubby legs that allowed this creature to manoeuvre across the ocean floor out of harm's way. So effective were this animal's defences that worms with legs lived on, despite the intense competition of the Cambrian seas, gradually evolving into creatures that began to experiment with a new life on land.

Just as Armstrong's leap into the annals of human history was made possible by the Moon, so too were the first land creatures dependent on lunar forces to help them take their small, but critical, steps on to dry land. Tides that relentlessly pulled the ancient seas back and forth from the ocean's coasts created a zone that allowed small, stubby-legged worms slowly to adapt to life beyond the seas. About 480 million years ago, descendants of the heavily defended Hallucigenia were in the process of evolving into the ancestors of today's terrestrial velvet worms, adapting to life on land, which was then an almost predator-free world, allowing them to abandon their heavy body armour.

Extreme changes of environment, such as the move out of the water and on to dry land, have often caused, over generations, some of the most dramatic evolutionary transformations in the history of life on Earth. These successful pioneers of the land developed small tubes (trachea) to let oxygen in the air pass deep inside their worm-like bodies. A system for males to pass sperm directly into the bodies of females ensured sexual reproductive success in an environment where there was no surrounding water to act as a freely available transport medium for fertilization. With their bodies divided into segments, each equipped with a pair of well-adapted legs, some of these creatures are thought to have transitioned into millipedes, centipedes and eventually, through the merging of various segments, into the familiar tripartite body plan of insects, with their endless variety. Genomic fusion and hybridization, thanks to successful chance matings between distantly related species, is how some experts account for the velvet-worm-like larval caterpillar phases of today's butterflies, wasps and scorpion flies.

Today's brightly coloured velvet worms are regarded as living fossils, almost exactly resembling their 400-million-year-old ancestors. Species such as *Peripatopsis alba* (white cave velvet worm) and *Peripatoides indigo* are shy creatures that live in warm, humid caves, forests and grasslands, mostly in the southern hemisphere. They conform almost exactly to the fossils of those creatures that first pioneered the original passage on to land, hundreds of millions of years ago. Once their ancestors had established their land-based niche some adapted into species that suited other terrestrial habitats. Others continued to live apparently without need for further evolution, preserving their genetic shape, form and function thanks to the preservative power of sexual reproduction.

More than 150 velvet worm species have been recorded to date, divided into two main families, of which the most ancient is called Peripatopsidae. Today many specimens are on the verge of extinction due to the destruction of rainforests and the draining of wetlands caused by the growth of human conurbations. Only time will tell whether the descendants of those ancient pioneers who first wriggled on to land can re-ignite their appetite for evolution, in order to save their ancient lineage.

Sea Scorpion

CLASS: EURYPTERIDA
SPECIES: JAEKELOPTERUS RHENANIAE
RANK: 78

*On the origins of an ancient instinct that became crucial
to the survival of many different species*

A DEEP-SEATED INSTINCTIVE fear of spiders is one of the most common of all human phobias. Like the development of physical forms, such as bones, shells or teeth, behavioural instincts have evolutionary roots reaching deep into the geological past. Charles Darwin devoted an entire chapter in his *On the Origin of Species* to the question of how certain inherited instincts – like those that cause a bee to build perfect hexagonal cells or a cuckoo to lay its eggs in other birds' nests – have variously improved the chances of a species surviving in the struggle for existence. Such instincts, said Darwin, have been refined and passed on through countless generations.

The instinct in some species to flee from danger probably reaches all the way back to when the oceans were populated by extremely scary relatives of today's arachnids (spiders) called sea scorpions. That these were the largest arthropods ever to have lived is now beyond dispute following the discovery in November 2007 of a huge sea scorpion claw from rocks in a quarry near Prüm, in the mountainous Schneifel region of Germany. By *c.* 400 million years ago, sea scorpions had taken over as the 'king-pins' of the marine eco-system following the extinction of creatures such as the Cambrian Anomalocaris. The enormous flesh-ripping claw, found by German palaeontologist Markus Poschmann, belonged to a species of scorpion called *Jaekelopterus rhenaniae*. This creature grew larger than a human – to more than 2.5 metres in length.

Sea scorpions (also known as eurypterids) fed off fish, life's original vertebrates, some of which were the ancestors of land animals, such as amphibians, reptiles and mammals, including man (see On Fish that Came Ashore, page 105). Eurypterid fossils have been found all over the world, indicating that not only were these prehistory's largest arthropod species but that they were also among the most successful. Sea scorpions spent their lives in shallow coastal areas, hunting bony fish, trilobites (see page 58) and other animals using stealth. Two pairs of eyes (similar in design to those found in today's spiders) helped these creatures burrow under the muddy sea floor, where they could lie in wait, keeping at least one eye out for any tasty morsels that happened to pass by. If near enough, the scorpion would launch itself at high speed, snatching at the unsuspecting prey, ripping it to shreds with its vicious claws. Its victims usually succumbed through complete surprise.

But not always, which is where the benefit of instinctive flight comes in. Those fish with the quickest reactions survived such fearsome predators, passing on their escapology skills to future generations. Thus a highly effective instinct for flight away from such sea monsters was triggered by the slightest sign of attack. Could a link between today's human instinctive fear of arachnids (spiders) originate with sea scorpions hundreds of millions of years before people first evolved? Arachnophobia is common enough: almost 80 per cent of twenty-five to thirty-four-year-olds questioned in a UK survey owned up to having some form of fear of spiders.

Increased levels of oxygen in the atmosphere, as a result of the land being colonized by plants and trees may have been one reason why scorpions of this era reached such massive proportions. Another reason may have been the lack of other large predators in the seas, allowing these scorpions to increase in size to the maximum limit that the physics of the arthropod design would permit.

Thanks to the evolutionary success of the instinct of flight, vertebrate bony fish eventually bit back in the form of their own larger species, such as placoderms – the first jawed fishes – and sharks (see page 71). These creatures successfully began to populate the seas from about 390 million years ago, which coincides with when the largest sea scorpions start to disappear from the fossil record. Eurypterids finally became extinct in the Great Permian Extinction 252 million years ago (see page 116), when as many as 96 per cent of all marine species were wiped out as a result of cataclysmic environmental changes. Only their closest arthropod relatives, the arachnids, survived this traumatic period, eventually becoming one of the most numerous arthropod groups on Earth (see page 441).

Unlike symbiosis, viral infection or selective pressures, instincts are forces for conservation and preservation, not adaptation and change. Just think: so conservative are the inhcrited instincts of humans that most of us will baulk at the sight of a harmless spider yet mindlessly risk picking up a loaded gun in a shooting range or wielding a sharp-bladed knife in the kitchen. Prehistoric sea scorpions were life's most powerful triggers for the evolution of instinctive fear in many ancient marine species and were therefore architects in the evolution of escape by fright. Such powerful inherited instincts are essentials to the success and survival of many of life's most resilient die-hards.

Sea Squirt

FAMILY: BOTRYLLIDAE
SPECIES: BOTRYLLUS SCHLOSSERI
RANK: 12

*Ancient sea creatures that could hold the keys
to improved human health*

MOLECULAR BIOLOGISTS ARE often regarded as today's best bet for extending individual human lifespans. Recent research into countering the natural ageing process has been focused on finding ways to re-grow or repair vital human organs – such as the heart, lungs or liver – that may be diseased or suffer from old age. Such a process of 're-vitalization' could go a long way to extending individual human lives.

A small but highly significant marine species whose family first evolved in the late Cambrian Period (*c.* 500 million years ago) and whose descendants, it is thought, eventually led to the evolution of man himself, may hold the key to how this miraculous process could one day be accomplished. Tunicates (more commonly known as sea squirts) are small jelly-like animals that usually glue themselves to the seabed in the luxurious environment of coral reefs (see page 52). They filter water through gill-like slits in their bodies, searching for microscopic food in the form of plankton (mostly algae and bacteria). Beneath the apparently simple body structure of these translucent, cucumber-like creatures lies a more complex system in which their vital organs (throat, heart, digestive tract, neural and reproductive systems) are protected by a layer of special connective tissue called mesenchyme, the same as that found in the embryos of humans – their latter-day descendants.

Sea squirt larvae are significant in evolutionary history as the first creatures known to have developed a notochord – a flexible, rod-shaped structure that in human embryos grows into a backbone (vertebrae). All vertebrates (which include fish, amphibians, reptiles and mammals) have backbones essential for activities such as swimming, walking, running and flying. Starting with the ancestors of sea squirts the chance change that split animal body designs into invertebrates and vertebrates began to reveal its true significance.

Most sea squirts use their flexible notochord only in the earliest stages of development when they swim – a bit like tadpoles – to a new location far away from their parents to reduce competition for food. They then take root, growing either alone or into a new colony, abandoning their tails, along with their precious notochords, to lead a sedentary lifestyle attached firmly to a reef.

One tunicate species, *Botryllus schlosseri*, has caught the attention of biologists because it is somehow able to spawn new colonies directly from its blood vessels rather than from its sexual organs – a process called vascular budding. Researchers at Stanford University, California, have found that a complete new tunicate colony can be formed from scratch via this process after successive generations of 'trial and error', especially if these creatures' sexual organs are removed. Efforts are now under way to decipher the genetic code of this species to find out how to switch this asexual regenerative process on or off, chiefly to see if it could be activated in humans. Such technology could allow people to 're-grow' damaged organs such as kidneys, a liver or even limbs. The California team's optimism stems at least partly from the fact that humans are distant descendants of sea squirts and share many genetic traits.

Another family of sea squirts, the aptly named Larvacea, never attach themselves to the sea floor. Instead they stay embryo-like all their lives, using their notochord-powered tails as they swim freely in the oceans, foraging for food. It has recently been discovered that these creatures may play a significant part in the overall health of the deep-ocean ecosystem. One example is *Bathochordaeus*, a widely distributed five-centimetre-long sea squirt that builds what are known as underwater 'houses', up to one metre wide, made out of mucus. The sea squirts use their notochord tails to beat water through these membrane-like structures to capture tiny particles of food. Once their mucus construction gets clogged up, they then abandon it, and swim off to build a new one – a process that occurs about once every four hours. The old mucus house then descends like a parachute, providing much needed nutrients to creatures that live on the sea floor.

Such house-building habits are now thought to have a quantifiable impact on the global climate. Researchers at Monterey Bay Aquarium Research Institute (MBARI), California, who have been measuring the effect of Larvacea on the oceans, believe substantial quantities of carbon dioxide are removed from the atmosphere as a result of this parachuting process, which pumps dead organic matter (which is partly made out of dissolved carbon originating from the air) to the sea floor. Their research is currently helping oceanographers and climatologists puzzle out what impact a rise in global sea temperatures may have on this natural biological process.

Despite being relatively inconspicuous and little known, sea squirts have certainly had a major impact on the planet, life and people in the deep evolutionary past as well as on deep-sea life in the present. As for the future, their potential lies in unlocking the secrets of how it may one day be possible to re-grow human tissue. Gerald Weissmann, editor of the Federation of American Societies for Experimental Biology magazine, believes that research into the reproductive quirks of sea squirts is '*a landmark in regenerative medicine . . . The biological equivalent to turning a sow's ear into a silk purse and back again.*'

Shark

CLASS: CHONDRICHTHYES
SPECIES: CARCHARODON MEGALODON
RANK: 38

Pioneers of adaptive immunity

CREATURES THAT RANK as one of life's greatest success stories do not automatically become the focus of human admiration and respect. Several of nature's most brilliant, robust and enduring designs are, in the eyes of humans, abhorrent organisms ranging from pests such as ticks (see page 415) and mosquitoes (see page 130) to monstrous, brutal killers.

Sharks are fish with jaws that evolved about 420 million years ago. Mostly they eat other fish or microscopic plankton and have no specific appetite for man. Their survival from their first forms (such as Stethacanthus) has remained almost unchanged for 400 million years. The key to their success is a combination of evolutionary innovation and an ongoing series of experiments and adaptations that has made these creatures among nature's most robust.

The evolutionary transformation of five-centimetre free-swimming sea-squirt-like marine creatures into massive ten-metre-long sharks took place over about one hundred million years. No one knows for sure where and how jaws first appeared. They may have arisen from members of bony-headed fish called Ostracoderms. Placoderms were the first true-jawed fish, some species growing to enormous sizes owing to their ability to devour just about anything and everything in their path. One type, Dunkleosteus, had a body ten metres long and a bite more powerful than *Tyrannosaurus rex* (see page 125). But placoderm supremacy was curtailed by a series of mass extinctions *c.* 360 million years ago and an evolutionary sister group took over as top marine predator. These were the sharks, creatures that have maintained their position at the head of the food chain ever since.

Three factors contributed to the sharks' long-term success. Firstly, sharks are among the most diverse animals when it comes to mastering the art of sexual reproduction. Some sharks reproduce by internal fertilization (copulating), others are oviparous, laying eggs with leathery cases commonly called mermaid's purses. Yet others lay eggs which females keep safely inside their bodies until they hatch. Another group behaves more like mammals, producing live young that feed on a special type of milk secreted inside the uterus. A final group indulges in the gruesome practice of oophagy where the largest youngsters hatch inside the mother shark only to eat their smaller siblings. As a result, only the strongest offspring survive. Such reproductive versatility is good insurance against mass extinction.

A second success factor is size. Legendary species include the sixteen-metre-long forty-eight tonne *Carcharodon megalodon*, a giant whale-killing shark that first appeared about sixteen million years ago. Its mouth could open as wide as two metres, revealing a set of gargantuan teeth, some of which were over twenty-two centimetres long. Environmental changes finally brought about the death of this monster of the deep. As the ice ages increased their grip, about two million years ago, fatty whales migrated to live in the icy polar seas, leading to the demise of the *megalodon* through lack of food. However, other shark species remain successful today – over 400 million years after their ancestral origins – thanks to a third innovation: the extraordinarily powerful system of adaptive immunity.

No one has yet compiled a comprehensive history for the origins of adaptive immunity or properly explained why this complex system, in which specialist immune cells remember previous infections in order to ward off repeat attacks, occurs only in vertebrate animals. One theory is that about 400 million years ago repeated swarms of various viral infections colonized the ancestors of today's sharks. Many viruses infect their hosts without causing them harm. Indeed they rely on the health of their hosts as viral replication vehicles. It is therefore in the survival interests of these infections to try to *prevent* other viruses from invading the same host organism. Over time, it is thought, tactics used by certain viruses to forestall infection by others may have helped establish the adaptive immune system that all vertebrates employ today.

Sharks display substantially greater levels of immunity from viral infection than other groups of living species – including plants, insects, bony fish, reptiles and mammals (including humans). Indeed, shark liver oil is today marketed as an 'immune system booster'. Norwegian fishermen in the eighteenth century regularly applied oil extracted from the livers of deep-sea sharks to open wounds as a healing agent. Meanwhile, Swedish researchers in the 1960s confirmed that the same immune-boosting compounds found in sharks are also abundant in human breast milk.

Mean, cruel and brutal are some ways of describing these highly successful creatures of the deep. Innovative, versatile and adaptive are others. In terms of impact, sharks have survived all five major prehistoric mass extinctions, although how well they will survive the human-induced sixth remains to be seen.

5

On Pioneers of the Land

How mutual collaboration between trailblazing species helped life colonize the Earth's barren landscape.

Prototaxites

> FAMILY: (UNKNOWN)
> SPECIES: PROTOTAXITES LOGANII
> RANK: 52

A fungal giant that paved the way for life on land

A MUSHROOM THAT'S as high as a house sounds like something out of *Alice in Wonderland*. But no. This giant fungus dominated the Earth's early terrestrial skyline for at least seventy million years, eventually becoming extinct about 350 million years ago. Its fossils still stud the deserts of Saudi Arabia, jutting out from dry rocks like the stumps of trees.

For a long while biologists believed this ancient form of land-life was indeed a tree – an ancestral conifer, perhaps? At least that was the view of John Dawson, a Canadian naturalist, who was the first to describe these log-like fossils back in 1857. He named them *Prototaxites* (meaning first yew) because of his understandable assumption that such large structures could only be the remains of ancient woodland. Other scientists were not so sure. Some said it was a giant stinging algae. The mystery seemed to be resolved in 2001 when, after extensive research by Francis Heuber from the Museum of Natural History in Washington, it was suggested that these were fossils of a giant fungus. Even more recently, in 2007, researchers at the University of Chicago put the matter beyond doubt. They discovered an uneven ratio between different types of carbon (isotopes) in the fossils, indicating that this was an organism that fed on organic matter (a heterotroph) and so was not a photo-synthesizing plant (an autotroph).

On Pioneers of the Land

Confirmation that *Prototaxites* were fungi underpins the huge impact these organisms must have had in the critical transition of life from sea to land. Even now it is hard to imagine a fungus so vast that when it wanted to reproduce, its mushroom-like fruit punched up skywards taller than a two-storey building. In the Devonian world, 420 million years ago, plants had yet to evolve their own proper roots. It was only through symbiotic relationships with fungi like *Prototaxites* that they could survive on land at all. Hairs at the bottom of plant stems connected with the branching hyphae of *Prototaxites*, which then delivered supplies of underground nutrients and water in return for food.

No estimates have been made as to how far underground a species like *Prototaxites* may have spread, but it is fair to assume that the hairs of each organism could have extended for many miles, providing underwater plumbing services to countless dependent plants. The impact on the Earth itself must have been equally profound. Feeding on subterranean bacteria, its network of digestive hairs (mycelium) would have broken up the hard ground, transforming inhospitable rock into crumbly soil filled with nutrients vital for other life-forms. The creation of such soils meant that over time plants were able to reach underground unaided to find their own supplies of water and minerals, anchoring themselves more firmly with deeper roots and so reducing their dependency on fungi. However, the fact that most plants today still rely on some kind of fungal association to survive is indicative of the huge ecological contribution made by pioneers like *Prototaxites*.

Many extinct species have played an important role in the history of evolution. Some contributions were so powerful that, like points on a railway track, they sent evolution down an entirely new path. Ironically, such paths often prove more advantageous for other species, sometimes driving the evolutionary protagonist into extinction (see also *Lystrosaurus*, page 115). That's what happened to *Prototaxites*. Once tree forests began to cover the land, these mighty mushrooms simply couldn't compete. Small invertebrate arthropods, such as worms, spiders, mites and millipedes, that lived on the increasingly fertile soils, now found these giants all too good to eat. As the forests grew, space for twelve-metre-high fungal fruits was at a premium. Only those fungi that adapted in size and developed new defences survived. These are the fungi we are familiar with today.

Exactly which group of fungi living today are the direct descendants of *Prototaxites* is not yet established, but the pioneering impact of these early land giants has few, if any, parallels.

Rhyniophytes

FAMILY: RHYNIACEAE
SPECIES: RHYNIA GWYNNE-VAUGHANII
RANK: 83

Extinct weeds that were the forefathers of trees

FANCY A WORLD without weeds? While the prospect may sound attractive to golfers and gardeners (see grass, page 381), it is only thanks to the curiosity of an early-twentieth-century Scottish doctor and amateur geologist that we now know what an extraordinary role ancient weedy plants played in the story of life on Earth.

One morning in the spring of 1912 Dr William Mackie spotted some unusual-looking rocks in a dry-stone wall surrounding fields near his home just north of Rhynie, a small village in Aberdeenshire, Scotland. Being a curious sort, he took them home to study in more detail under his microscope. He was amazed to discover they contained an intricate cross-section of beautifully preserved fossils, showing highly detailed systems of what looked like plant stems and cells. Nearly 400 million years ago Rhynie was nothing like Scotland is today. Located somewhere near the equator, this was a geothermic spa-land steaming with boiling pools of bubbling mud. In this hot, sweltering place, giant geysers would frequently spout huge fountains of scorching water laced with silicon from deep beneath the earth. When this silicate water landed on nearby vegetation and plants it killed them instantly and quickly cooled, turning their petrified remains into perfect fossils of stone. The fossils of Rhynie have since shown scientists exactly how primitive plants began one of history's most remarkable transformations: from floating seaweeds to woody shrubs and ultimately into enormous trees that could colonize the land many thousands of kilometres away from the water's edge.

The plants' 'trick' was to become 'vascular'. Mosses are land plants that have no roots, leaves or stems and can survive only in damp habitats close to water. Leafy plants and trees today use a design for their survival derived from these early pioneers at Rhynie. For plants (as for animals) life on land is dramatically different from living in the seas. For a start, land-life has to grapple with the force of gravity, which has far less effect in water. Learning to grow upright – against the force of gravity – was no simple adaptation. The plants at Rhynie show the first attempts at a solution with the development of a dead cellular structure called lignin in which layer upon layer of tightly packed cells are stacked and interwoven to make rigid stems. These structures are what ultimately led to the construction of that sturdy constituent of shrubs and trees – wood. Lignin has a host of other essential functions too. Even in dry conditions, it supports plants so they can stay upright as close to the Sun as possible. Lignin also toughens stems to support tubes (xylem and phloem) that can transport nutrients and water up and down the plant. Vascular systems stop plants drying out and allow them to grow tall and support leaves – essential prerequisites for the eventual evolution of most of the world's trees.

Living on land had other hazards such as the Sun's ultraviolet radiation, which can easily destroy the delicate structures within a living cell's DNA molecules (and can lead to skin cancer in humans). Early plants, like Rhyniophytes, developed ultraviolet-resistant spores that protected their offspring from the Sun's radiation even if germination was delayed by a lack of sufficient water. Such compounds have been common to all land plants since.

If Rhyniophytes were solely responsible for the development of all the technology needed to make trees, then their position in our top one hundred ranking would have been a great deal higher. But

On Pioneers of the Land

Rhyniophytes mostly used scales on their stems, leaving it to their descendants, the Euphyllophytes (meaning 'good leaf plants', such as ferns, see page 86) to experiment with the concept of hanging green solar panels, in the form of leaves, from their stems to better catch the sunlight. Rhyniophytes were also limited by having roots that grew only along the ground, rather than burrowing underneath, restricting their growth to a maximum height of about fifty centimetres. Underground roots that supply plants with water and nutrients and prevent trees from toppling over came with the Lycophytes, the world's first true trees (see page 86).

Rhyniophytes were seedless weeds. Their spores had to land on wet ground for successful germination, a limitation that became more significant as the Earth's continents converged into a single landmass, Pangaea, c. 300 million years ago and dried out.

Ultimately, the significance of these early vascular plants was in starting the long but crucial evolutionary process of adapting to life on land from ground-hugging mosses into tall, majestic trees – a process that took about forty million years. A world without trees would have severely limited the potential for all land-life, including the evolution of primates such as monkeys and humans. Also, the atmosphere would never have benefited from the enormous capacity of the rainforests to absorb carbon dioxide, a process which has helped to keep the Earth cool and make it better suited for life on land.

It was also largely due to the emergence of early vascular plants that oxygen levels rose high enough to entice larger forms of sea-life to leave the water and to evolve air-breathing lungs (see lobe-finned fish, page 106). Indeed, apart from worms and insects, most animal life would have remained limited to the seas were it not for the food and oxygen provided by these early land-dwellers.

Lepidodendron

> FAMILY: LEPIDODENDRACEAE
> SPECIES: LEPIDODENDRON SP.
> RANK: 11

Gigantic, fast-growing trees that fuelled the industrial world

GO TO GLASGOW and there, in the south-west corner of the Scottish city's Victoria Park, is a small insignificant-looking Victorian museum. Inside, however, the building opens out to reveal one of Britain's most precious and surprising geological treasures – a group of eleven massive tree stumps, each at least a metre wide, with roots that bulge and buckle across the excavated ground. These are naturally occurring, fossilized sandstone casts from a once vast 330-million-year-old tropical forest that stretched all the way from Norway in northern Europe to Pennsylvania in the United States of America.

The trees from which these stumps were forged were once the most prolific on Earth. They are club mosses, of which only a few species survive today. The biggest and most dominant of all were long extinct *Lepidodendron* trees that grew up to two metres wide and a staggering fifty metres high.

The *Lepidodendron* was like no tree alive today. One peculiarity was that it grew extraordinarily fast, reaching its enormous height within just ten to fifteen years. For most of its life it bolted upwards like a giant telegraph pole – without branches or true leaves. Green photosynthesizing scales grew on the outside of its trunk, which branched outwards at the bottom forming networks of gnarled roots to provide structural stability and support. Only towards the end of its life did this curious tree sprout a canopy of branches, and then they appeared only at the very top of its trunk. Clusters of small green photosynthesizing blades hung from this late-growing canopy, and, unlike today's trees, these land pioneers reproduced only once in a lifetime. Thousands of small, light cones filled with spores fell from the tips of its branches like a swansong before the tree died. Mostly these landed close by, although some would have been blown farther away by the wind, littering the forest floor with seedlings that would begin the cycle of rapid growth all over again.

At least half of all the trees in the forests that sprang up during the Carboniferous Period (360–299 million years ago) are thought to have been club mosses (Lycophytes), of which by far the most numerous was the *Lepidodendron*. Since they produced no branches or leaves until the very end of their lives, these were trees that could grow extremely close together without suffering from a lack of light. The result was that the forests they formed were the most dense of all time. It is thought that up to *2,000 trees* could grow in a single hectare of soil. Fast growth, enormous height and short lifespans meant these ancient forest floors were rapidly choked with layer upon layer of dead wood. Over time, the enormous belt of woodland that thrived in the balmy tropical climate across Pangaea was piled high with so much dead wood that no amount of bacteria, insects or fungi could hope to rot the lot. As rapid Carboniferous Period climate changes brought frequent flooding and rising and falling sea levels, so the wood sank into the swampy forest floor, each layer of trees being buried deeper and deeper under the ground until . . . their remains were discovered just a few thousands years ago by humans.

Not that Bronze Age Britons (*c.* 3000–2000 BC) had the slightest inkling that this magic black rock originally came from enormous trees that inhabited 300-million-year-old forests. Only in the relatively recent past, when countries like Britain were in the throes of a chronic shortage of wood, did people first start to understand that the fuel they were burning came from the fossilized remains of the Earth's ancient forests. Coal is the remains of ancient trees (ranging in age from fifty million to 350 million years old) buried in water, bogs and sediment that were starved of oxygen, causing the natural process of rotting to stop. Over enormous tracts of geological time, these non-rotted hulks of wood were transformed by pressure, heat and chemical decomposition into the black stuff we call coal. Carbon from the air, used by these trees to build their woody trunks, was therefore buried underground, gradually reducing levels of carbon dioxide in the atmosphere. Over time, this helped to cool the climate, a contributing factor in the numerous ice ages that have occurred over the last three million years.

The impact of the world's coal deposits on the planet, life and people has been truly transformational. The deployment of coal in Britain during the eighteenth century to power industrial machines (chiefly steam engines) solved an energy crisis that had come about from chopping down too many trees to make ships and houses, and to clear fields and farmland. With the success of high-pressure steam, pioneered by Cornishman Richard Trevithick in 1801, came humanity's first ever form of fully independent power, freeing this species from the limitations of alternatives such as animals, water and wind. Soon everyone wanted to enjoy the benefits of industrialization. Within a generation coal was being mined in huge quantities in all inhabited continents of the world in a frenzy to feed a surge of mechanical modernization.

Without coal there would have been no industrial revolution – at least not when it occurred, in the early eighteenth century. Although oil is now the world's fossil fuel of choice (since the invention of the internal combustion engine *c.* 1880) as much as 40 per cent of the world's electricity is still powered by coal today. Also, coal (in its processed form, coke) is essential for the extraction of iron from iron ore. Everything made from iron and its derivative steel that has ultimately come to shape the modern world – from knives and forks to skyscrapers – exists thanks to these deep deposits of what were once the Earth's most ancient trees.

Despite their great success on land, *Lepidodendron* eventually became extinct because of competition from other better-adapted types of trees. Seeds, broad leaves and the recruitment of other creatures (such as insects) to aid their sexual reproduction were just some of the innovations made by other species that eventually led to the *Lepidodendron*'s demise. Yet few plant species have had quite such a profound impact.

Azolla

> FAMILY: AZOLLACEAE
> SPECIES: AZOLLA PINNATA
> RANK: 60

A prolific fern that caused the Earth's climate to flip from a greenhouse to an icebox

FEW SPECIES IN our survey of those that have had the greatest impact on the planet, life and people have as bizarre a claim to supremacy as the small water-dependent fern called *Azolla*. Its influence pivots on a dramatic period in the history of the world's climate. Ferns were the first plants to develop true leaves, called megaphylls. These broad green solar panels provide a large surface area for photosynthesis, giving this family a significant survival advantage over more 'primitive' plants, such as Lycophytes, that used microphylls – scales and thin hair-like structures on their stems, trunks and branches (such as *Lepidodendron*, see page 82).

About 20,000 species of ferns exist today, ranging from common woodland bracken to the more exotic *Cyathea*, a genus of tree-ferns that grow up to twenty-four metres high in the forests of Australia and New Zealand. Other, less conspicuous species still hug the water, harking back to their evolutionary origins. Ferns, unlike conifers and flowering plants, lack flowers or seeds. Instead, like fungi, they produce spores that rely on moisture for successful germination. Fertilization also requires a fern's sperm-like reproductive cells to swim through water. The story of how one genus of tiny floating water ferns, called *Azolla*, managed to have quite such a big impact on all other life begins about forty-nine million years ago.

On Pioneers of the Land

At that time in the Earth's history, about sixteen million years after the mass extinction event that wiped out the dinosaurs (see page 126), the Arctic polar region, familiar to us today as a continent-sized ice pack, was a balmy semi-tropical swamp. Alligators and snakes slithered through the stagnant water, and giant mosquitoes grew as large as a human head. Research from an Arctic expedition has revealed that for about one million years the Earth's temperature rose dramatically, owing to abnormally high quantities of carbon dioxide in the air. What caused levels of this greenhouse gas to rise to as much as 3,500 parts per million (compared to 330 p.p.m. today) is still unknown. A sudden increase in volcanic activity perhaps? Or maybe the mysterious release of millions of tonnes of methane (another greenhouse gas) normally trapped as frozen deposits on the sea floor? Whatever its cause, the effect was dramatic. Samples (cores) taken from deep down under the seabed show how average temperatures at the North Pole soared to about twenty-three degrees Celsius. Because of the way the Earth's continents were arranged at that time, the Arctic sea was almost completely isolated from the rest of the world's oceans and so had few or no deep-water currents. River water poured into the northern seas following an increase in rainfall, which formed a layer of low-density fresh water on top of the high-density salty sea.

Conditions for the water fern *Azolla* to make its mark could not have been better. Like legumes (see *Rhizobia*, page 30) but unlike most other plants, *Azolla* had established a symbiotic partnership with nitrogen-fixing organisms, in this case cyanobacteria. These bacteria live inside the leaf cells of the fern, turning nitrogen from the air into a nitrate solution that the plant can use for growth. In return, the fern feeds the bacteria sugars derived from its photosynthesis. So powerful is this partnership that *Azolla* is able to grow extraordinarily fast – it's been dubbed a 'super-plant' that can more than double its mass every two to three days. About forty-nine million years ago was a great time to be alive for *Azolla* plants. With abundant supplies of carbon dioxide in the air, an internal source of nitrates, a favourable climate and plenty of nutrients in the water, nothing could stop this fern from runaway reproduction.

And for about 800,000 years nothing did.

Arctic cores show that a massive four-million-square-kilometre swamp of *Azolla* grew right over the top of the world and as far south as Britain. In the process, it sucked up so much atmospheric carbon dioxide that levels fell steeply to 650 parts per million, reducing air and sea temperatures and flipping the world from a greenhouse to an icebox. Average temperatures have been cooling ever since, providing a better environment for the globe-trotting habits of mammals, the spread of grasslands and conditions that eventually encouraged apes to come down from the trees and experiment with walking on two feet (see *Australopithecus*, page 184).

The climatic contributions of this group of ferns are set to continue well into the future. As polar ice melts from the injection of ever-increasing levels of carbon dioxide into the atmosphere by modern man, geologists are now discovering what may have happened to the remains of that enormous prehistoric swamp. When the ferns died, they sank to the salty depths of the sea, where a lack of oxygen prevented their proper organic decomposition. As layers of sediment piled on top, this dead organic matter is thought to have become compressed and crushed over millions of years into what may be vast fields of yet-to-be-discovered crude oil and natural gas.

As the polar ice melts, exploration of the Arctic seabed and drilling for cores has become big business. Russian submarines, Norwegian ice-ships and the Canadian military are all preparing to make new territorial claims in anticipation of a dash for the huge reservoir of oil trapped at the roof of the world. If future generations of humans do find it and then pump it out and burn it, releasing the trapped carbon into the atmosphere, maybe these small ferns will be needed again? Not that their efforts are likely to provide a quick-fix solution to the current problems of global warming. Judging from their previous efforts, mopping up is a process that takes the best part of one million years.

Norway Spruce

> **FAMILY: PINACEAE**
> **SPECIES: PICEA ABIES**
> **RANK: 22**

A highly celebrated species that boasts today's oldest living individual

WHILE CHRISTMAS COMMEMORATES the birth of Jesus Christ more than 2,000 years ago, the origin of the Norway spruce, pulled out of the ground, taken indoors and decorated as part of this annual festival, is a great deal older. Researchers from Umeå University in eastern Sweden recently discovered a single specimen of Norway spruce (*Picea abies*) in the Swedish mountains that they say has been alive for more than 9,500 years. This venerable tree has been nicknamed Old Tjikko and may be the oldest-known living organism in the world. The key to its longevity is simple. Whenever the main trunk dies (typically about every 600 years), its ancient roots simply sprout a new one in its place. No need for fertilization, no need to spread seeds.

Old Tjikko belongs to a group of trees called conifers, which evolved after the club mosses (see page 82) but long before flowering plants. Between 300 million and one hundred million years ago, these trees were the predominant forms of plant-life on land. They out-competed other trees, such as the telegraph-pole-like *Lepidodendron* (see page 82), with a range of survival skills, of which Old Tjikko's cloning trick is but one. With the rapid rise of flowering trees and plants, particularly in the last hundred million years, conifers have finally been beaten into retreat. Except when deliberately cultivated by humans their domain is usually restricted to the world's least hospitable climates – or to soils that are so thin and sandy that few other types of tree can effectively compete.

A range of evolutionary innovations accounts for the conifers' historic success. Their use of seeds instead of spores allowed these trees to spread inland, far away from the water's edge. They developed branches that grow outwards all the way up their trunks, and not just from the top, allowing them to absorb greater quantities of light, especially important for survival in higher latitudes where the Sun stays lower in the skies. Perhaps the biggest impact of conifer trees is down to their ingenious method of self-defence – resin. This thick sticky liquid slowly oozes down the sides of the tree's branches and trunk, plugging up holes bored by insects. As it solidifies, it protects the tree against fungal and bacterial infection.

Amber fossils (formed from resin deposits) date back 345 million years, although those containing the remains of ancient insects are common only from about 150 million years ago. Today's most prolific amber deposits are found in the Baltic, a region where conifer trees continue to dominate the landscape despite the region's cool climate and the impact of various ice ages in the last three million years. Resin has had its own special influence on human culture. Its sticky viscous qualities are perfect for use as a natural glue and varnish for woodworking, incense for religious ceremonies, and for rosin, a material which helps 'sharpen' a string player's bow.

The success of conifers is manifest from their wide variety, ranging from pines, firs and spruces to cypresses, junipers and redwoods. They include the tallest (*Sequoia redwood*), oldest (*Norwegian spruce*) and broadest (*Ahuehuete cypress*) living species in the world. In their heyday, conifers were as ubiquitous in the tropics as they are in today's northern latitudes – as evidenced in coal strata laid down over hundreds of millions of years. The flattened shape of the juniper is living proof of a tree that has adapted to environments where light shines directly from above, as it does in the tropics, maximizing the area available for leaves to gather light.

Other conifers include the world's most massive trees. The giant redwoods of North America have developed immunity to what, for other trees, would be the greatest threat of all – fire. Redwoods, like some pines in Mexico, have become so used to forest conflagrations that they cannot properly regenerate without them. Their seeds spring loose from their cones only in the extreme heat of a fire. Such events clear the forests, providing sufficient light for new redwood seedlings successfully to germinate.

For more than 200 million years conifers eclipsed all other trees in the forest – club mosses, ferns, cycads – even pushing some of them into extinction (for example the *Lepidodendron*). Their success stemmed from a combination of good design, unfussy habits and a range of innovative systems for self-preservation.

Without man, however, the long-term survival of the Norway spruce and other conifer species (and many of its cousins – the pines, firs, larches, cedars and hemlocks) looks a lot less secure. Cloning may have worked brilliantly for Old Tjikko, but it is not a strategy that bodes well for future generations. Diversity is the surest protection against climate change or viral infection. The lack of species variety among the conifers of northern latitudes today is a symptom of this once great family's gradual demise. To find out why, skip on to the rise of flowering plants and trees (see page 142) . . .

Earthworm

FAMILY: LUMBRICIDAE
SPECIES: LUMBRICUS TERRESTRIS
RANK: 1

*Subterranean wrigglers whose constant
burrowing fertilizes the soil*

ACCORDING TO CHARLES Darwin, no living thing has had such a profound impact on history as the earthworm. So fascinated was he by these humble creatures that he devoted an entire book, published in 1881, to the formation of soil (then called 'vegetable mould') through the action of worms:

> *The plough is one of the most ancient and most valuable of man's inventions; but long before it existed, the land was in fact regularly ploughed, and still continues to be thus ploughed, by earthworms. It may be doubted whether there are many other animals which have played so important a part in the history of the world, as have these lowly organized creatures.*

Earthworms are annelids, a phylum of creatures whose evolutionary past stretches back at least to the Cambrian Explosion *c.* 530 million years ago, when the trilobites first developed sight and marine creatures evolved bones and shells. *Burgessochaeta* is an ancestral, twenty-segmented sea worm whose fossils were found by Charles Walcott in the Burgess Shale.

Descendants of these marine creatures came ashore at the time of the first invertebrate invasions of the land, *c.* 450 million years ago, making their living in damp soils broken up by bacteria, fungi and the roots of colonizing plants. These earthworms have been ploughing up the earth, ventilating the soil and nourishing terrestrial ecosystems with their excrement ever since. Five mass extinctions have occurred over the last 500 million years, some of which devastated up to 96 per cent of all marine species and 70 per cent of all land species, but none of them ever touched these creatures. Slice a worm in half and it re-grows as if nothing happened. Divide one half and the same thing happens. One worm even survived forty such butcherings, all in the name of science.

The effects of worms on human history are as profound as they are unwritten. French scientist-cum-poet Andre Voisin was one of the few experts who properly highlighted the role of worms in the birth of ancient human civilizations. Were it not for their continuous regeneration of soils around damp river valleys such as the Nile, Indus and Euphrates, early agricultural societies in Egypt, India and Mesopotamia could never have succeeded in building humanity's first large-scale urban communities. Even the Egyptian pyramids, said Voisin, were built thanks to the nourishment of the soil by earthworms. It was only because of their hard work that farmers could take time off from tilling the soil themselves to work as a labour force for their pharaoh's ambitious building projects.

Throughout human history earthworms have unintentionally but undeniably triggered the rise of civilizations. Wherever earthworms plough, people thrive. When worms perish, societies collapse. Infertile soils led to the demise of the people of ancient Sumeria. Rising levels of salt as a result of irrigating the land with sea water killed off the worms around the mouth of the Euphrates river and the soil turned sour. By 2000 BC their civilization was so weak from lack of food that they fell easy prey to Assyrian invaders from the north.

It might be easy to think that worms matter little today, replaced by artificial fertilizers and pesticides that guarantee soil fertility anywhere and everywhere they are spread. But no. Once again it was largely thanks to the earthworm that the unsustainable nature of using such methods was originally exposed.

Rachel Carson was a teacher and environmental campaigner of the 1950s and 1960s. She is famous for warning that Americans might one day wake up to discover they could no longer hear the birds singing in the trees. The reason, she said, was that artificial pesticides, such as DDT, were poisoning the soil. While robust earthworms are able to tolerate such toxins, for those creatures that eat worms it is a different story. As few as eleven worms that have ingested DDT are enough to poison a robin – either killing it or making it sterile. Since robins regularly eat up to twelve worms an hour the use of DDT put their populations, along with those of similar birds, at risk of annihilation. The prospect was terrifying enough to have DDT banned in the USA by popular demand (although it is still used in some African countries, such as Kenya, to support the global cut-flower trade – see rose, page 354). Carson's book *Silent Spring* (first published in 1962) inspired the founding of the US Environmental Protection Agency in 1970. Therefore, the robust digestive system of the earthworms is, in a curious way, inextricably linked to the birth of the modern environmental movement.

But which worm species has had most impact? There are some 15,000 species of segmented worms in the annelid phylum, including leeches and marine polychaetes as well as earthworms. They range from the now rare but enormous purple-headed Giant Gippsland (*Megascolides australis*), a native Australian worm that grows up to three metres long, to the extremely common red wrigglers (*Eisenia foetida*), vermicultural alchemists that turn kitchen vegetable scraps into rich garden compost. *Lumbricus terrestris*, the European earthworm, is now probably the most prolific and invasive species in the world. Its success is largely thanks to the spread of Europeans from *c.* 1600 onwards. Immigrant farmers inadvertently brought these earthworms, sometimes called 'night-crawlers', to the Americas in everything from the soil in their potted plants and their horses' hooves, to the treads of their boots and the wheels of their wagons. Today, there is hardly a region of North America where Europe's earthworms have not made a home for themselves. There they continue to plough, ventilate and fertilize the soil to the general benefit of life in and on the Earth.

But their presence is not always welcome. Artificially introduced invasive species almost always have a darker story to tell (see also eucalyptus, page 271, and rabbit, page 344). Native American redwood forests that were once free of worms are now being invaded by this European species. It is relentlessly munching its way through carpets of fallen leaves – a vital habitat for native American insects, amphibians and ground-dwelling birds. Without leaf litter on the soil's surface, the seedlings on which the future of these forests depends will not germinate. Within a generation or two, some experts fear that those forests not already mined by man may instead be undermined by the worms he has introduced.

Darwin's fascination with these ubiquitous noiseless digesters of the Earth was well placed, confirming their rank as our Number One species (see table on page 436). Whatever they lack in glamour, colour or a sense of adventure (most worms move only about fifty metres in their four-year lifetime), they make up for in their constant ploughing, harrowing, fertilizing and recycling of that most precious of all the planet's assets – the living earth itself.

Dung Beetle

> **FAMILY: SCARABAEIDAE**
> **SPECIES: SCARABAEUS SACER**
> **RANK: 68**

A venerated recycler from the most diverse group of animals on Earth

BEETLES COME IN a greater variety of shapes and forms than any other order of creatures. They comprise more than *a quarter of all known animal species*. At least 350,000 types have been described to date but between three million and eight million species are thought to exist in total. These are nature's most varied group of multicellular creatures and have endlessly fascinated man.

Why so many species? The answer may lie in a branch of evolution that has only recently come under scrutiny – hidden perhaps because it runs in a contrary direction to the competitive state of nature emphasized by Darwin in his *On the Origin of Species*. Co-evolution is best described as an intimate evolutionary dance between different types of living things in which a spiral of mutually beneficial adaptations triggers a diversification into numerous slightly different species.

Beetles were the original pollinators of plants and trees, more than 100 million years before the emergence of birds, bees or moths (see pages 122, 157 and 266). Beetles pioneered a mutually beneficial relationship with the first seed-bearing trees, the cycads (see page 98), which evolved about 280 million years ago. Until then, the sexual reproduction of plants and trees was a mostly hit-and-miss affair. Wind power was the only viable alternative transport medium for plants on land. Aquatic plants used water currents (as green algae and mosses still do today) to carry male genes (pollen) to female eggs. But in a world far away from water, wind was all there was. By 280 million years ago, a new mechanism became essential as the world's landmasses converged into the one large super-continent of Pangaea, where rainfall patterns shifted and inland areas became drier than ever.

Successful plants and trees were those that produced enormous clouds of pollen – the more pollen, the greater the chance that the wind would blow at least one or two particles towards their intended targets. But for any plant, producing so much pollen was expensive and wasteful. All power to those species that found an alternative means of transportation for their precious male genes. Enter the beetle. Even the earliest fossils show how well these creatures adapted to hot arid environments. Folding wing designs allowed them to squeeze into narrow places, inside the cracks in bark or under rocks, where larger fixed-wing predators could not penetrate (see dragonfly, page 101). This knack of retreating into tight corners served these creatures well in the harsh inland environments of Pangaea.

Cycads were trees that developed their own innovative response to the challenges of life on a dry super-continent (a few species survive today). Producing seeds allowed their offspring to germinate without the need for constant damp conditions, as required by spores. Seeds contain a store of food, packed around the embryo and surrounded by a tough outer coat designed to survive the heat. A second innovation was the ancient cycads' production of scents that attracted beetles to feast off their pollen and seeds. As the beetles roamed around neighbouring trees, they spread the cycads' pollen – a far more efficient and targeted form of transport for the trees' male genes than having to rely on their pollen blowing about willy-nilly in the wind (see also honey bee, page 157).

As cycad species spread across the globe so did subtle changes in the aromas they produced, each species evolving its own special scent. As clusters of similarly inclined beetles found themselves attracted to the same types of trees, their gene pools became localized – or isolated – in the process, with each separate community evolving into its own distinct species as successful characteristics become predominant.

How humans have become so familiar with such a wide variety of species is largely thanks to the efforts of European beetle hunters. One Russian collection, begun by Tsar Peter the Great (ruled 1682– 1725), runs to more than *six million individual specimens*, each one separately housed, mounted, listed and labelled. The fascination for beetles in Darwin's day among British Victorian naturalists was just as pronounced. On one occasion Darwin was out hunting for beetles and was holding a specimen in each hand when he saw another, even rarer example. His eminently practical response was to put one of the beetles he was holding into his mouth for safekeeping so he could retrieve the third. In the process he made yet another scientific discovery – in confined spaces some beetles eject a noxious poison . . .

How to pick out a single representative from the 350,000 or so known species living today presents another interesting challenge. One candidate, venerated since the beginnings of recorded human history, is the scarab or dung beetle, a creature that feasts off the excrement of larger animals. The scarab's preference is for the dung of herbivores such as elephants, camels, cows and sheep. So enthusiastic are these creatures for their share of the cake that in some parts of the world thousands of beetles converge within moments of a fresh helping landing on the ground. Some Australian dung beetles (a genus called *Onthophagus*) have even developed a knack of clinging to the hairs around a wallaby's anus, just to make sure they are first in line when it comes to fresh deliveries.

Scarab beetles knead dung into spherical balls that they roll across the ground to a suitable place for burial, providing a rich store of food. Female scarabs lay their eggs inside these balls, giving their young larvae a source of protection as well as nutrition. Ever since large terrestrial herbivores began to make their mark, starting in the age of dinosaurs, the ecological contribution of these assiduous recyclers has been immense. Even today it is estimated that the US cattle industry alone saves $380 million a year thanks to the burying of livestock faeces by scarabs and other insects.

The ancient Egyptians venerated one species, *Scarabaeus sacer*, as a living god on Earth because of its dung-rolling antics, recorded in many scarab-shaped amulets (funereal brooches). Pictures of scarabs were also painted or etched on to the walls of ancient Egyptians' tombs (see plate 4, bottom).

Dragonfly

> **ORDER: PROTODONATA**
> **SPECIES: MEGANEUROPSIS PERMIANA**
> **RANK: 50**

*A fearsome fixed-winged insect that once
dominated life on land from the skies*

AGGRESSION IS AN instinct most often found in species at
the apex of the 'pyramid of life' – a natural arrangement of creatures
that occurs in all ecosystems. Light from the Sun is absorbed by
plants and turned into sugars. Herbivores (creatures who feed on
plants) stand to gain only about 10 per cent of the Sun's trapped
energy, since the rest is used by the plants for their own growth
and reproduction. Further up the pyramid, the number of species
and the size of populations gets smaller and competition for survival
increases. Carnivores (creatures that feed on other animals) stand
to gain only about 10 per cent of the energy trapped in herbivores.
Therefore, in their bid to survive, these species have to be the most
aggressive of all.

Many of the most spectacular examples of the survival of the
strongest over the weakest (one of Darwin's recurrent themes)
come from species at the top of this pyramid. Over the course
of evolutionary history, the fortunes of the 'super-species' at the
top have risen and fallen like stocks and shares today, sometimes
descending into a niche or occasionally going bust and sinking
into extinction.

Among the land's earliest animal king-pins were the dragonflies.
The aggression (and success) of ancient dragonflies arose partly from

the fact that these insects were among the first to take to the skies. High levels of oxygen in the Carboniferous Period's atmosphere and the increasing height of plants as they morphed into the Earth's first trees are likely to have assisted insects' first flight. It is not too difficult to imagine how it may have happened, even if no fossil evidence has yet been found of insect 'proto-wings'. Flaps originally designed to regulate body temperature could have gradually adapted into wings and some daredevils probably experimented with gliding from tall plants (or trees) to the ground rather than crawling all the way back down. Over time, muscles developed to assist the gliding process, itself made easier by the dense atmosphere, which contained greater quantities of atmospheric oxygen, derived from the forestation of vast tracts of formerly barren land. Higher oxygen levels also help to explain why the ancestors of today's dragonflies were able to grow quite so large – powering them with the extra energy needed for growth and flight.

Griffinflies, an extinct order related to today's dragonflies, grew as large as eagles. Swooping down from the skies with a wingspan of more than half a metre, these creatures preyed on unsuspecting terrestrial animals. Compound eyes, perhaps descended from those of the trilobites (see page 58), reached their zenith in flies like these. Typically the lenses featured more than 30,000 individual facets, providing a near 360-degree view of the world.

Such huge flies could either indulge in a small, juicy four-legged vertebrate or alternatively devour less powerful insects. A major impact of their intense predation was the development of the folding wing in other insects (a characteristic of the group Polyneoptera – see dung beetle, page 97). The design, thought to have evolved as a defence against dragonflies, allowed vulnerable insects to creep under logs or into crevices and cracks in rocks to escape becoming a fixed-wing predator's prey. It is therefore thanks to dragonflies and their ilk that the highly successful escapologist instincts and skills of house-flies, beetles, wasps, bees, ants and other flying folding-wing insects ultimately evolved – a process that began from the later Carboniferous Period onwards.

For as long as fifty million years (*c.* 320–270 million years ago) the skies were ruled by giant dragonflies (such as *Meganeuropsis permiana*) and griffinflies (*Meganeura moni*), the largest insects ever known. But being big came at a price, making them more susceptible to sudden environmental change. Declining oxygen levels probably contributed to the demise of these enormous creatures. Insects do not have lungs. Instead they rely on oxygen diffusing through their bodies via small tubes called trachea. Large flyers therefore depend on high concentrations of oxygen so the life-giving gas can properly permeate throughout their body. When oxygen levels fall, the maximum possible size for flying insects falls too.

Today's smaller dragonflies thrive nonetheless. They express their powerful inherited instincts for survival in at least two remarkable ways. First, a dragonfly's genius at acrobatic flight usually means it can use its near 360-degree vision to catch its hapless victim while still on the wing. Spine-like studs jut out from its legs forming a cage-like grip from which there is no hope of escape. Some flies take their meals back to their roosts, others prefer to consume their catch fresh while still aloft.

A second highly aggressive survival strategy is sexual. Charles Darwin wrote about 'sexual selection' as a force of variation between genders of the same species in his *On the Origin of Species*. The sexual antics of dragonflies offer an extreme example of sexual competition between males (for other examples, see *Archaeopteryx*, page 121, and sperm whale, page 171). Having identified a compatible mate (colour and flight patterns act as species hallmarks), a male dragonfly will grasp a female by the neck in mid-flight and hold her there for as long as it takes to transfer his sperm successfully (see plate 3, bottom). The two of them will form what is known as a 'copulating wheel', the male not releasing his grip until the female yields, even if it takes several hours. In a unique evolutionary adaptation, the male's genitals are designed to scoop out any sperm already inside the female that may have been placed there during a previous liaison. The dominant male will then attempt to hold the female hostage until she has laid her eggs, a final bid to ensure other males of the same species do not repeat the trick he performed. In such a manner do genes from the strongest, most dominant individuals get passed on, via inheritance, to future generations. Such tactics are nature's surest way to cultivate a predatorial master-race.

6

On Fish that Came Ashore

How the descendants of bony fish clambered ashore and were jerry-wrenched into a diverse range of forms.

Lobe-finned Fish

CLASS: SARCOPTERYGII
SPECIES: PANDERICHTHYS RHOMBOIDES
RANK: 51

A semi-aquatic fish that began the process
of adapting to life on land

TODAY, LITTLE ELSE causes as much confusion and debate (and often fury) in science than the preoccupation with biological reclassification, ever since evolutionary descent rather than outward form became the accepted new order. Willi Hennig (1913–76), a German biologist, first formalized the quest to reveal life's true family tree and originated today's system of putting species into groupings or branches, called 'clades', based on evolutionary descent.

One of the biggest problems is that as more data comes to light from sequencing the genomes of different species, the more it seems that evolution works convergently – that is, when ecological opportunities arise, unrelated organisms often end up evolving in similar ways. One of the most bizarre consequences is that it is now incontrovertibly clear that humans belong to a clade of creatures called Osteichthyes – the bony fish. In fact, in cladistic terms, all land creatures – from birds, cats and cows to elephants, mice and humans – are technically divisions within the world of bony fish. So, if land animals are descended from bony fish, whichever species were the first to make it possible for vertebrates to clamber ashore must have been of enormous evolutionary significance.

Air-breathing apparatus, prototype arms, legs, fingers and toes

– all vital equipment for moving about in a gravity-dominated environment – originally evolved from the four lobed fins of a class of fish called the Sarcopterygii, common in the Devonian seas. The full potential for exploring a new life beyond the seas reached its zenith with lobe-finned fish such as *Panderichthys rhomboides*, now long extinct, which lived about 380 million years ago. Fossil evidence shows that this species pioneered opportunities for feeding beyond the water's edge. Because of the low levels of oxygen in the shallow waters near the coasts at that time, the most successful of these fish were those that could absorb small quantities of oxygen by gasping in air through their mouths. At the same time, pushing a long fish-like body along the shallow sea floor using rear fins made perfect sense for a creature that cruised along the coast, its eyes emerging from beneath the waves – hunting and prowling for fresh food – half fish, half crocodile. *Panderichthys* fossils clearly show how bones in its rear (pelvic) fins helped it rise out of the water. By pushing these fins against the ground it could then drag its body along like a modern, walking catfish.

Several elements critical to the success of animal life on land came together in this environment, about 375 million years ago. Bones such as the humerus, radius and ulna can be traced back to lobe-finned fish, as can several joints that have since become limbs, wrists and ankles in land animals. Tetrapods, terrestrial four-legged descendants of lobe-finned fish, soon began to populate the seashores. Lines of immature seal-like creatures called *Ichthyostegas* sheltered from aquatic predators by seeking the comparative safety and protection of the mostly untrodden land.

Lobe-fins were, for a long while, among the most dominant fish in the seas. One huge genus, *Hyneria*, was the king fish-killer of its era. This fast-swimming two-tonne, four-metre-long carnivore roamed close to the shore, using its powerful front fins as legs to haul itself along the sea floor, searching for fish stranded by the tide.

Of course, a bony spinal column, air-breathing lungs and four jointed limbs aren't the only requirements for success on land. Sophisticated senses such as hearing and even the capacity for making sounds – the ancient origins of birdsong and music – are now thought to have first evolved in bony fish.

Tiktaalik

CLASS: TETRAPODOMORPHA
SPECIES: TIKTAALIK ROSEAE
RANK: 88

Caught in the act: the flat-headed half fish, half land animal that boasted the world's first neck

EVERY SO OFTEN the discovery of an entirely new species fills in a conspicuous gap in the jigsaw puzzle of life. The most prized discoveries are known to science as 'transition fossils'. These are the imprints of creatures that show the process of evolution during a clear transition from one group of living things to another. Such creatures provide snapshots of the evolutionary process at work. One of the most famous examples is *Archaeopteryx* – half reptile, half bird (see page 121).

Experts who discover a transition fossil tend to get rather excited – which is exactly what happened to two American fossil-hunters, Neil Schubin and Steve Gatsey, who went out camping on one of the most remote islands on Earth in the summer of 2004. The pair had already spent six long years on a quest to find the missing evolutionary link between bony fish and four-legged land creatures, called tetrapods. The challenge they set themselves was to find a species that could clearly be identified as a genuine half-fish, half-tetrapod pioneer. If Darwin's theory that all creatures are descended from common ancestors was correct and if, as other fossil finds indicate, all life began in the seas, then somewhere in the rocks there should be fossils that show the transition from a purely sea species to one that could thrive on land.

The Sun never sets in the summer at Ellesmere Island, in northern

Canada. But in the winter it is dark twenty-four hours a day. Yet somewhere, in an arc stretching about 1,500 kilometres wide, were rocks that Schubin and Gatsey reckoned were just the right age for their transition fossils. They knew roughly what they were looking for. The buried remains of the creature they sought were probably not more than about a metre long and at least 370 million years old. All they had to do was dig. After spending four summers over the previous six years on the same quest, Schubin struck fossil hunter's gold:

'*I cracked the ice and saw something I will never forget: a patch of scales unlike anything else we had yet seen in the quarry. This patch led to another blob covered by ice. It looked like a set of jaws. They were, however, unlike the jaws of any fish I had ever seen. They looked as if they might have been connected to a flat head . . .'*

One day later, and after some rather delicate hammering, the unmistakable snout of a second mysterious beast could be seen poking out of the rocks. Gatsey spent the rest of the summer recovering the fossilized creature, bit by bit, so it could be taken back to the laboratory for analysis.

The creature has been called *Tiktaalik*, the Inuit word for large, freshwater fish. It had scales on its back like a fish, but, like a land animal, it also featured a flat head and neck. Most amazingly, its fins were composed of bones that correspond to those found in all land animals. These were the upper arm, forearm and parts of the hand with their corresponding joints that in later creatures became shoulders, elbows and wrists. The conclusion that the pair drew was incontrovertible. Here was a genuine 375-million-year-old transition species showing the evolutionary link between lobe-finned fish and four-legged, land-vertebrate tetrapods.

So what of the impact of a creature such as this? No one can be sure if the adaptations seen in *Tiktaalik* are unique, since who knows what other secrets lie hidden in the icy rocks? What's more, other creatures such as *Eusthenopteron* and *Acanthostega* have also been discovered that show evolutionary transitions away from the seas and on to the land. But *Tiktaalik* is special because it is the oldest creature found so far whose fins clearly show the arrangement of bones that in later land animals would become wrists, palms and fingers. *Tiktaalik* was a creature that could do the world's first push-up – or, in the words of Schubin, it was able to '*drop and give us twenty*'.

Any time you bend your wrist back and forth, or open and close your hand, you are using joints that first appeared in the fins of the half-fish, half-amphibious creature called *Tiktaalik*. That's its place, and therefore its significance pretty much speaks for itself.

Dimetrodon

ORDER: PELYCOSAURIA
SPECIES: DIMETRODON GRANDIS
RANK: 82

The sail-back pelycosaur that found a way of keeping itself at just the right temperature

JERRY MACDONALD IS a detective who has spent the last forty years exclusively investigating footprints. Old ones. Very old ones. In fact, the prints that MacDonald specializes in are typically about 270 million years old. MacDonald's scene of investigation is also fairly eccentric. High up in the mountains of New Mexico he has removed more than one hundred tonnes of rock, revealing a series of ancient footprints that tell their own fascinating story of lives once lived. MacDonald then takes them away for storage in the nearby town of Las Cruces, where they are stacked, catalogued and sometimes sent off to laboratories or museums for further analysis.

This is 'Pelycosaur Heaven' – the name of MacDonald's fossil trackway quarry in the mountains of New Mexico. Its name comes from its record of footprints belonging to a group of ancient creatures, the most successful land animals of their age.

Dimetrodon, a member of the pelycosaurs, grew up to three metres long, making it the largest reptile of its day. First appearing about 286 million years ago, this genus was one of the prehistoric world's longest survivors. For more than thirty million years *Dimetrodon* was a top predator on land. Unlike amphibians, its waterproof skin held moisture inside its body, allowing it to migrate far inland and, if necessary, to survive in hot, arid conditions. And like all reptiles,

On Fish that Came Ashore

Dimetrodon laid its eggs in watertight shells, freeing itself from the highly competitive environment near the coasts where amphibious creatures were forced to breed.

But *Dimetrodon* is most famous for the dramatic-looking sail on its back. This ingenious adaptation worked like a radiator, allowing its body to heat up quickly in the morning sun. Packed full of blood vessels, this sail meant *Dimetrodon* was ready to hunt at least an hour before non-sailbacked reptiles, a big reason for its long period of success. In the evening it could turn its sail away from the sun, allowing heat to escape, ensuring its body always remained as close as possible to its operational optimum.

These traits are also common in mammals, such as humans. Our body automatically keeps to the same temperature, within a few degrees, night and day, regardless of conditions outside. Reptiles do not. Their behaviour is limited by their need to warm up in the sun. *Dimetrodon*, therefore, represents the first significant milestone along the path of evolution from reptiles to mammals. *Dimetrodon* has also left a distinct impression thanks to its teeth. Indeed, *Dimetrodon* literally means 'two teeth'. These were the first reptiles known to chew their prey, mashing it up into pieces before swallowing (unlike other reptiles that gulped down great chunks of food whole). Different types of teeth of varying sizes made this possible. Some were adapted for tearing up flesh, others for chewing it. This gave these carnivorous creatures an advantage because it takes less time to digest and draw energy from pre-processed, chewed-up food. As with the ability to regulate body temperature, their varied teeth and eating habits were features that later expressed themselves in *Dimetrodon*'s mammal-like descendants.

MacDonald's footprint investigations leave no doubt as to the success of these impressive beasts, which emerged a good fifty million years before the first dinosaurs. While fossils often reveal what an animal may have looked like, they do not generally say so much about how it behaved. Footprints, however, offer a snapshot of what animals were actually doing at a given instant in time. They are long, unbroken trails of prehistoric walking, darting, swerving and pouncing. The *Dimetrodon* tracks from the mountains of New Mexico were submerged by mud and silt along what was once the shoreline of a vast inland sea. They show that invertebrate animals and small shellfish were frequently exposed by the receding tide, leaving large predators, like the *Dimetrodon*, to feed on the pickings.

After thirty to forty million years of uninterrupted success, the *Dimetrodon* suddenly died out during the Great Permian Extinction (see page 116). It did not disappear through bad design, however, or even because of the rise of a rival species. These creatures were swept away by powerful geological forces, leaving a void at the apex of the pyramid of life, without which a new group of reptilian dinosaurs might never have arisen to take their place . . .

Lystrosaurus

> ORDER: THERAPSIDA
> SPECIES: LYSTROSAURUS OVICEPS
> RANK: 66

The barrel-chested herbivore that survived when all around it perished

I FEEL A SENSE of impending doom as we approach the story of *Lystrosaurus*. It isn't helped by the front-page headline in my morning newspaper: 'The Methane Time-Bomb' tells the story of Dr Örjan Gustaffason, leader of a marine research project studying methane emissions from beneath the Arctic sea. His team from the University of Sweden has discovered a series of 'chimneys' on the sea floor from which large undissolved bubbles of methane gas are steadily discharging. In some places, says the report, concentrations of methane are escaping from the sea floor at more than one hundred times background levels.

What seems to be happening beneath the seas today is frighteningly similar to what is thought to have happened much earlier in Earth's history. About 252 million years ago vast quantities of this gas started bubbling up to the ocean surface. It was released from huge frozen methane deposits (called hydrates) under the sea floor that had been accumulating over millions of years through the bacterial decay of organic matter. The slow grinding together of the Earth's tectonic plates to form the super-continent Pangaea may have caused a large enough increase in global temperatures (thanks to carbon dioxide spewing out of volcanic vents) to have provided the trigger.

What on Earth Evolved? . . . in Brief

The eruption of giant bubbles of methane was bad news for life on Earth, leading to the largest mass extinction of all time. Up to 96 per cent of all sea creatures and as many as 70 per cent of all land species perished in the Great Permian Extinction. Nothing as appalling has happened to life on Earth since or before – not even the extinction that killed off the dinosaurs 65.5 million years ago. Not only is methane about twenty times more potent as a greenhouse gas than carbon dioxide, it also reacts with atmospheric oxygen to make carbon dioxide and water. About 252 million years ago this chemical reaction caused global levels of atmospheric oxygen to plummet from 30 per cent to just 12 per cent over the course of about 20,000 years – the merest fraction of geological time.

Suddenly having to cope with an atmospheric oxygen content of 12 per cent is equivalent to living at the top of Mount Everest today. Large reptiles that depended on meat for food never had a chance, which spelled disaster for the thirty-million-year reign of *Dimetrodon* (see page 112). Giant griffinflies the size of birds (see page 102) were also at a hopeless disadvantage, dependent as they were on high concentrations of atmospheric oxygen to power their over-sized bodies.

Plants suffered too, especially those in forests. One of the tell-tale signs of death by suffocation is the absence of coal deposits. The 'coal gap', dated to between about 250 and 243 million years ago, is explained by a decline in oxygen levels that forced peat-forming plants into extinction. It wasn't until several million years later, when new species of plants developed better tolerance to lower oxygen levels, that the coal-making process could begin afresh. Only those creatures with lucky adaptations that made them less susceptible to low levels of oxygen, and those that didn't depend on a supply of fresh meat, could thrive.

Enter *Lystrosaurus*. Judging by the fossil record, for at least two million years after the Great Permian Extinction, this metre-long herbivorous creature – half-hippopotamus, half-pig in appearance

– represented about half of all vertebrate animal life on land. So numerous were they that their remains have been found in the rocks of every continent, once again confirming that 250 million years ago all the world's continents were indeed joined together into the super-giant Pangaea.

The success of *Lystrosaurus* may have been down to its barrel-like chest, which had an especially large breathing cavity, putting it in a good position to survive the plummeting oxygen levels. It used its two horn-like canine teeth to burrow into the ground. Perhaps its subterranean habits explain this creature's large lung capacity, a happy adaptation for survival in stale, low-oxygen underground air. *Lystrosaurus* gained a survival advantage in being able to regulate its body temperature (see also *Dimetrodon*, page 112). Being vegetarian was good news too. With the extinction of so many other species, there was just enough plant food left for herbivores like *Lystrosaurus* to survive.

The consequences of *Lystrosaurus*'s survival through the Great Permian Extinction were immense because, without it, the line of warm-blooded vertebrates that eventually evolved into mammals would probably have died out 252 million years ago, along with *Dimetrodon* and all the rest. Yet, following this most traumatic of evolutionary episodes, the baton of supremacy among land creatures passed to a different group, one that led to the dinosaurs. The descendants of *Lystrosaurus* retreated into the margins of terrestrial society, becoming smaller and furrier, and limiting themselves to hunting mostly at night. Then events once again took evolution in a new direction with another extinction event, which unfolded just as dramatically 65.5 million years ago.

Quetzalcoatlus

> ORDER: PTEROSAURIA
> SPECIES: QUETZALCOATLUS NORTHROPI
> RANK: 49

*The world's largest flying creature and the
first global circumnavigator*

DREAMS OF HUMAN flight reach far back into history to
the myths of ancient Greece. But this human fancy was preceded
in reality by that of other vertebrates that took to the skies aided
by nothing more than the power of biological evolution some 230
million years ago. What's more, these animals mastered the power of
flight a good seventy-five million years *before* even the world's most
primitive birds could be heard singing their first morning chorus
(see *Archaeopteryx*, page 121).

Pterosaurs were reptiles that evolved alongside the dinosaurs, soon
after the Great Permian Extinction 252 million years ago. Their
bodies were adapted for flight by the growth of an extraordinarily
long and strong fourth finger, which protruded like a strut to support
the front edge of a giant webbed wing. Exactly which species first
took to the skies and why is unknown. Fossils of small crow-sized
creatures called *Sharovipteryx* with flaps of skin stretched between
their fore and hind limbs have been found in Kazakhstan. Reptiles
like these were well adapted for escaping predation by leaping off
hills, stones, cliffs and ledges. Over time, their success probably led
to the first species of pterodactyls, which could actually fly and not
just glide. Exactly how pterodactyls were capable of powering themselves into the air, and not just gracefully gliding through it, was

demonstrated by American inventor Paul MacCready. In the 1980s this engineer rose to the challenge of building a half-size working model of the largest flying reptile that ever lived.

Quetzalcoatlus northropi was a genuine giant. Fossils found in Texas suggest its wingspan stretched to as much as twelve metres – larger than many small aircraft today. This awesome creature, which lived between eighty-four and sixty-five million years ago, was a master-piece of aeronautical engineering, reaching what experts believe are the maximum limits possible for any living creature – it was the Howard Hughes *Spruce Goose* flying boat of its day.

And, as MacCready's successful simulations demonstrated, this behemoth really did fly. The half-size model, launched from a trol-ley in the Mojave Desert, California, on 27 January 1986, made six successful sorties using battery-powered electric motors instead of organically fed muscles. With wings flapping, looking just like a Cretaceous pterosaur, the model spiralled, dipped and swooped through turns as sharp as 270 degrees, re-creating the graceful winged magic of a long-lost age.

Pterodactyls were the first creatures to have global oversight of the world, capable of travelling from pole to pole. For the first time in the history of life, the entire planet could become a single creature's habitat. Large species, like *Quetzalcoatlus northropi*, could fly for thousands of miles at a stretch, connecting dispa-rate lands as the single continent of Pangaea continued to break apart to form the discrete continents we know today. Who knows what opportunistic species hitched a ride on this global transport system? What seeds, spores, bacteria or viruses were spread by this new form of long-haul flight? Birds would later reshape the world by spreading all manner of organisms, from influenza to fruit. Pterodactyls most likely did something similar, but many millions of years before.

Pterodactyls never made it through the catastrophic extinction that struck 65.5 million years ago, despite their enormous success over the previous 170 million years. This trauma killed off not only the dinosaurs but eight out of nine orders of birds (see page 126). Now only one order survives, providing life's single link with the era of vertebrate flight pioneered by pterodactyls so many aeons ago.

Archaeopteryx

> **CLASS: AVES**
> **SPECIES: ARCHAEOPTERYX LITHOGRAPHICA**
> **RANK: 41**

*A boastful show-off that spawned new flights
of evolutionary fancy*

IN HIS LATER years Charles Darwin became increasingly fascinated by sex. He wasn't interested in the salacious kind beloved of
today's tabloid newspapers and lifestyle magazines. His focus was
far more scientific. It was trained on the extraordinary differences in
appearance and behaviour between males and females of the same
species. Why are male gorillas so much larger than females? Why
are peacocks so colourful compared with peahens? Why do batteries
of male birds of paradise with elaborate plumage perform eccentric
dancing rituals in front of crowds of fussy females?

Darwin believed all these variations could be explained by a
second type of natural selection pressure. Sexual selection was not,
he said, a struggle for existence, but a competition *'between the males
for possession of the females'*.

Birds are creatures that have often followed this trend, as in the
case of peacocks and birds of paradise. Females of these species are
attracted to males that show off. They wish to mate only with the
most brightly coloured or best performing males since they instinctively reckon the chances of their children mating will be greater if
they inherit the best 'showing-off' characteristics. In this way the
selection pressure for genes responsible for brightly coloured feathers – or elaborate dancing rituals – in male birds becomes ever more

pronounced over time, as has happened in some bird species (for example, peacocks, parrots and drakes). The curious natural history of sexual discrimination is now thought to have provoked one of the most significant developments for life on Earth in the last 150 million years – the flight of the world's first bird (for another example of the effects of sexual selection, see *Homo erectus*, page 187).

Archaeopteryx is one of the most famous fossil discoveries of all time. The impression in rock of a half-reptile, half-bird creature caused a sensation when the fossil was dug out of a limestone quarry in Germany in 1861. Just two years after Darwin published his *On the Origin of Species*, here was a fossil that showed a creature in an evolutionary transformation. Like *Tiktaalik* (see page 109), *Archaeopteryx* was (and still is) one of the most stunning transition fossils ever discovered. Its skeleton was unmistakably reptilian but it had the feathers of a bird, even though they jutted out from each vertebra along the length of its tail, and not in a clump from the final vertebra, as in modern birds. Its discovery turned up the heat in the nineteenth-century debate between the Creationists and the Darwinists.

Today theories as to the origin of modern birds and the place of *Archaeopteryx* have themselves evolved. Traditionally it has always been thought that the definition of a bird was a creature (usually able to fly) that is covered in feathers. But the amazing story of 150-million-year-old *Archaeopteryx*, the world's ancestral bird, has been complicated by the discovery in China in 1998 of (non-flying) dinosaur fossils showing the unmistakable imprints of feathers. If dinosaurs had feathers then perhaps birds are actually a branch of the dinosaurs? Most experts are now agreed that birds do represent the one and only surviving group of dinosaurs. They are descended from theropods (see page 125), fossils of which have recently been discovered also showing feathery features.

So, if feathers evolved *before* birds, what was their original purpose? Chinese dinosaur fossils show that theropods with feathers were not flying creatures because they had no wings, leaving experts to suppose that feathers in dinosaurs evolved first as a means of insulation, and were later adapted – as is so often the case in nature – by *Archaeopteryx* and its descendants for flight. But now another theory is gaining ground. It goes back to Darwin's acute observation of the major differences in looks and behaviour between the sexes in some species. Feathers are outgrowths from scales, a common feature of reptilian creatures. Some reptiles are renowned for their elaborate courtship rituals and displays, for example, the male komodo dragon.

Such dynamics may have led unwittingly to feathered flight in birds. Somewhere along the dinosaur-to-bird evolutionary highway, flashy scales that refract the light, producing pretty colours, proved irresistible to females. This could have caused a runaway evolutionary spiral that resulted in elaborate plumage as seen in *Archaeopteryx*. Once endowed with such a covering, the benefits of extra insulation and later the re-orientation of feathers for the aerodynamics necessary for flight were adaptations no more remarkable than the jaw bones of fish receding in the skull, amplifying vibrations in the air allowing terrestrial animals to hear.

But what caused the feathers of *Archaeopteryx* to become adapted into wings that supported flight? Its habitat was surrounded by lagoons and sparse vegetation no more than three metres high, suggesting a ground-up origin of flight. Like its theropod relatives, ancestors of *Archaeopteryx* probably ran along the ground, occasionally leaping into the air to catch insects to eat. As they did so, the ones that flapped their feathers were increasingly successful, until eventually those with the largest, most aerodynamic arrangements spawned wings that supported take-off.

Archaeopteryx is the oldest-known relic of that branch of dinosaurs that survived the mass extinction 65.5 million years ago and went on to fill our skies with the songs of thrushes and skylarks we are familiar with today. The impact of bird flight, which co-existed with the soaring pterodactyls for about seventy-five million years, has been one of the most powerful forces of evolutionary change. Unlike pterodactyls, which ate fish and insects, many birds are herbivores. Such a flexible diet helped their long-term survival, enabling them to radiate into myriad modern forms.

Seeds wrapped in fruit, originally designed as large packets of fast food for roaming dinosaurs, evolved into smaller bribes (such as berries) that could be carried in the belly of a creature the size of a dove and deposited in its excrement many miles away from the plant's parental origin. The overpowering success of angiosperms (flowering and fruiting plants and trees) over other types of trees (such as conifers, cycads) in the last 150 million years is living proof of the incalculable effect birds have had on the shaping of the modern world. The story began with *Archaeopteryx*, the first known flying avian, whose appetite for sexual display seems to have led to an outcrop of feathers that were subsequently adapted for the purpose of better survival – jumping into the air – eventually turning them into the lords and ladies of the skies.

Tyrannosaurus

ORDER: THEROPODA
SPECIES: TYRANNOSAURUS REX
RANK: 34

*An iconic terrorist that forced other creatures
to find new forms of survival*

DINOSAURS, ALTHOUGH EXTINCT for more than sixty-five million years (apart from the birds, see page 121), have nevertheless had a profound impact on the planet that lives on today. The most successful dinosaur group of all was the theropods. Not only were these among the first dinosaurs to emerge, about 230 million years ago (*Eoraptor*) but they are the only surviving branch of dinosaurs to exist today. Birds, with their common theropod characteristics of wishbones and three-toed feet, are this group's living descendants. The rest disappeared in a giant mass extinction 65.5 million years ago – the same catastrophe that finished off all the other dinosaurs, as well as the pterosaurs (see page 118).

The most fearful theropod and the most famous dinosaur fossil was *Tyrannosaurus rex* – a gruesome creature that lived between seventy-five million and sixty-five million years ago in North America and Canada. Similar creatures, such as *Tarbosaurus*, roamed the forests of east Asia, while *Allosaurus* and *Gigantosaurus* lived in South America. Collectively, these dinosaurs would have created a climate of fear among other vertebrate animals during the Jurassic and Cretaceous Periods (206 million to 65.5 million years ago).

Anyone living close to Chicago can get a good first-hand impression of the sheer size and power of creatures like these by taking a trip to the city's Field Museum. Here the largest and most complete fossilized *T. rex* skeleton is to be found. She is named Sue, after Sue Hendrickson who discovered her in 1990 near the Cheyenne River Indian Reservation in western South Dakota. Sue clearly shows why the *Tyrannosaurus* was such a fearsome creature, with bone-crushing jaws more than 1.2 metres long and one metre wide. Its individual teeth were curved and serrated, many of them longer than a human hand, allowing the beast to crush its victims to death almost instantaneously, swallowing up to seventy kilograms of meat in one go. Key to these creatures' success was a fully improved stance that meant their entire weight was taken on their hind legs, allowing them to balance effectively on two feet and increasing their manoeuvrability despite their enormous body size (twelve metres long).

The devastating event that killed off all the dinosaurs 65.5 million years ago, including *T. rex*, was probably the chance impact of a ten-kilometre-wide meteorite. This cataclysm plunged the Earth's climate and atmosphere into a new dark age where luck, not evolutionary fitness, became the main criterion separating survival from extinction. Large terrestrial reptiles were most at risk – especially those that fed on flesh. As populations plummeted, food became scarce. If the actual impact or climate didn't kill off *T. rex*, then starvation soon finished the job. Yet long after their demise, the impact of theropods like *Tyrannosaurus* continues to reverberate in the shape, form and function of much of the biological world today.

Descendants of *Dimetrodon* (see page 112) and *Lystrosaurus* (see page 115) thrived by staying small and hiding in the woods and trees. These were little shrew-like mammals, which spent time protecting their young, and benefited from a covering of fur to keep their bodies warm so they could hunt at night when there was less threat of attack by theropod monsters. Fertilized eggs that grow *inside the mother's body* were another adaptation that ultimately evolved in many mammals as a direct response to the threat posed by dinosaurs (including birds and pterosaurs) that regularly helped themselves to eggs. Mammals were oviparous – they gave birth to live young. They also had mammary (milk-producing) glands, which meant females did not have to leave their nests to find food for their young until it was safe to do so. In a world dominated by giant meat-eaters like *T. rex*, mammary glands evolved as an essential component of the mammals' Jurassic survival kit.

7

On Biodiversity

How beauty, collaboration, deception, parasitism and vice wove terrestrial life into a rich carpet of countless species.

Mosquito

FAMILY: CULICIDAE
SPECIES: ANOPHELES GAMBIAE
RANK: 14

*Nature's female hypodermic injectors that
carry species-changing infections*

NEVER UNDERESTIMATE THE power of a parasite. Never misjudge the meanderings of a mosquito. And be especially fearful of females. They are the ones that bite.

Mosquitoes evolved at about the same time as the dinosaurs, 230 million years ago, and have probably had a bigger impact on vertebrate animals than any other living creature. Females puncture the skin of animals with nature's equivalent of a hypodermic syringe, sucking out the blood to use as a source of protein for producing eggs. Such behaviour has had far-reaching consequences for the evolution of life on land.

There is no desire in these bloodthirsty females to inflict damage or disease on the hosts that they feed on – far from it. It's just that mosquitoes are perfect vehicles for the transmission and replication of other types of parasitic life. All manner of microscopic parasites have found a way of spreading themselves among the land animal population by inserting themselves into the mosquito's saliva, thanks entirely to the power of a female mosquito's bite and flight. Males feed on nectar and honeydew.

Immediately after a female mosquito inserts her syringe into an animal, she injects a shot of saliva that contains anti-coagulants designed to prevent the host's blood from clotting too fast. It's via this saliva that other parasites – such as worms, bacteria, viruses and protozoa – are able to gain entry into the vertebrate bloodstream, spreading rapidly through the complex network of arteries and veins. Once inside, these parasites try to compromise their host's immune system so that they can successfully breed. Their offspring either spread via the host's guts and faeces, sometimes through body liquids such as breast milk, or they may lie in wait until another mosquito draws out its own small vial of blood. The parasite then infects the saliva of this new mosquito, preparing for when she bites another host – and so the transmission of the parasite species moves successfully along.

About 300 diseases that affect humans are transmitted via mosquitoes, including some of the most dramatic in history as well as many of the worst afflictions still affecting populations today. The most famous is malaria – so called because until 1894 it was thought to be caused by stale or 'bad air' (*mal'aria*) since the disease was frequently contracted in marshy, wetland areas. But when French physician Charles Laveran took blood from an infected patient and studied it under a microscope, he spotted a tiny amoeba-like creature moving around, powered by a flagellated (whip-like) tail. His deduction that this protozoan parasite, *Plasmodium*, was spread via mosquito bites earned him a Nobel Prize in 1907.

Laveran's discovery was momentous. Malaria is a disease that can cause all manner of symptoms from fever and flu to coma and death. Today approximately 515 million people have malaria worldwide. Deaths from the disease total from one million to three million people a year – most of them in Africa. Historically, this disease has been one of mankind's major killers, ranking alongside smallpox (see page 13) and influenza (see page 4), although its origin is at least ten times more ancient than most other diseases, which emerged in humans alongside the domestication of farm animals *c.* 10,000 years ago (see On Agriculture, page 201). Scientists are sure that humans have been affected by malaria for such a long time because they can trace a genetic mutation that has arisen in some people as a result of the disease. This mutation stretches back at least 5,000 generations and amply demonstrates the impact that mosquitoes, as flying vectors of disease, have had on human evolution.

Some people in Africa, which is where malaria has maintained its grip for the longest period of time, have developed genetic immunity. Their red blood cells have become flattened and rigid, a response that makes it hard for the *Plasmodium* parasite to attach itself and enter bodily organs. This is known as sickle cell trait. Although it brings immunity to disease, the genetic differentiation carries a high cost. Flattened blood cells are less efficient at carrying oxygen around the body and they cannot always squeeze through their blood system's narrowest passageways, called capillaries. Sickle cell trait symptoms are usually mild. But some individuals inherit a more serious form of the condition called sickle cell anaemia. This greatly shortens lifespans and can severely affect people's quality of life. But its origin as a genetic response to the *Plasmodium* parasite represents a classic example of how one species (a disease-carrying mosquito) can alter the evolutionary genetics of another species (humans) by acting as a conduit between a parasite (*Plasmodium*) and its host.

Other parasites have also latched on to the powerful transmission opportunities afforded by female mosquito saliva. Forms of yellow fever (caused by a virus of the family Flaviviridae) result in symptoms ranging from headache to coughing up of blood and internal bleeding, and may also result in death. Several appalling epidemics have been recorded in human history since the fall of the Roman empire, mostly in Africa, Europe and the Americas. Outbreaks have sometimes caused hundreds of thousands of deaths. In 1802 Napoleon sent an army across the Atlantic to suppress a rebellion in the Caribbean island of Haiti. His forces were decimated by yellow fever, as a result of which any lingering French aspirations to invade the recently liberated mainland of America via French-controlled Louisiana were finally curtailed. Consequently Haiti

Under attack: Viruses inject their own genetic material (in blue) into a bacterial cell to hijack its copying machinery.

1

Water Mould: Sporulating branches of *Phytophthora infestans* punching upwards through the stomata of an infected potato plant.

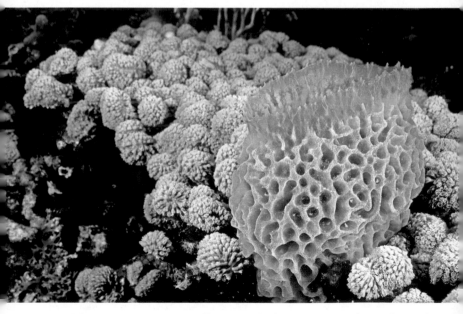

Sponge: The azure vase sponge (*Callyspongia plicifera*) growing on the sea floor off the Caribbean island of Grenada.

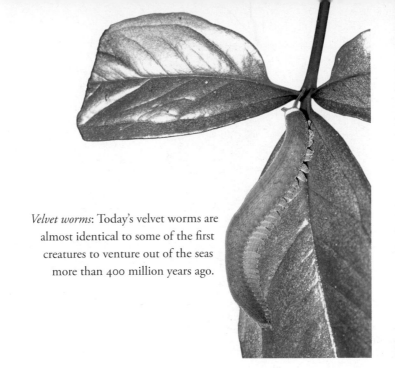

Velvet worms: Today's velvet worms are almost identical to some of the first creatures to venture out of the seas more than 400 million years ago.

Dragonfly: Checkmate: the mating rituals of the dragonflies are among the most aggressive sexual antics in nature.

3

Flea: Not 'Lord of the Flies', but 'Queen of the Fleas': Miriam Rothschild hard at work poring over parasites.

Earthworm: This cartoon was published in the British magazine *Punch* in 1881 – the same year Darwin brought out his book on earthworms.

Dung beetles: Scarab beetles were venerated in ancient Egypt, as shown in this cartouche, which signifies the Pharaoh Amenhotep II (reigned *c.* 1427-1400 BC).

Homo erectus: An artist's impression of *Homo erectus* based on 800,000-year-old-fossils found in Java, Indonesia.

Homo sapiens: Cave paintings of animals at Lascaux, France, created by prehistoric people to contact the spirit world.

Camel: Golden wonder: Mali King Mansa Musa sits on a golden throne and is approached by a traveller on a camel. Part of a Catalan Atlas dated to 1375.

Rubber tree: Henry Wickham employing Amazonian women to weave him enough baskets to transport rubber seeds back to Kew.

Rice: Planting out paddy fields in the Far East, from a nineteenth-century Japanese print.

Cannabis: Indian deity Shiva prepares to drink a cup of bhang, made from crushed cannabis flowers.

Cow: The sacred cow, depicted here as the mother of
the world, her body studded with deities.

– a land of enslaved African cotton labourers – became only the second European colony to gain sovereignty (after the recognition of American Independence in 1783).

Wuchereria bancrofti, a type of nematode roundworm (see page 55), also employs the services of nature's most prolific parasitic transporter, the mosquito. These worms can grow as long as six to eight centimetres and live inside the human lymphatic system, especially around the groin and legs. Elephantiasis, a highly unpleasant disease caused by these worms, has been known to cause a man's scrotum to swell to the size of a football. Despite the existence of effective treatments, an estimated forty million people worldwide suffer from its effects, especially in countries such as Ethiopia.

Charles Darwin believed that the 'struggle for existence', which underpinned his evolutionary theory, was driven by the power of overpopulation, which '*inevitably follows from the high rate at which all organic beings tend to increase . . . As more individuals are produced than can possibly survive, there must in every case be a struggle for existence*'. The disease-carrying mosquito subverts this basic Darwinian dictum by controlling population levels, sometimes before they get large enough to compete aggressively with other species. Therefore, those best adapted to withstand parasitic attack become those best placed to survive, as shown by the case of human sickle cell trait. As a result, at least a few individuals in a given species will have immunity to disease. Their survival potentially gives rise to new species (see also HIV/AIDS, page 9).

Diagrams that illustrate the web of life usually put producers (such as plants) at the bottom and carnivores (such as humans) at the top. That is wrong. Biting insects such as mosquitoes – see also fleas (page 134) and tsetse flies (page 138) – are the real king-pins of wild ecosystems through their constant infecting and modifying of the most powerful species through the transfer of disease-causing parasites. In the process they have shifted and shuffled endless evolutionary strands throughout the countless generations of life.

Flea

ORDER: SIPHONAPTERA
SPECIES: XENOPSYLLA CHEOPIS
RANK: 18

Spectacular jumping insects that have plagued human history

MIRIAM ROTHSCHILD (1908–2005) was one of the twentieth century's most brilliant naturalists (see plate 4, top left). She campaigned vigorously against all kinds of cruelty to animals and converted to being a strict (albeit eccentric) vegan, famous for wearing Moon boots in the winter, training shoes in the summer and white wellington boots in the evenings. But her most celebrated passion was the study of parasites – from flukes to cuckoos. She even found time for a certain rather rare worm *'which lives exclusively under the eyelids of the hippopotamus and feeds upon its tears'*. Among creatures with these lifestyles, her greatest fondness was saved for a species of tiny jumping insect, which is why she was often called 'Queen of the Fleas'.

Fleas were in the family. Miriam, like her father, was an avid collector. She spent thirty years cataloguing her father's world-famous collection of fleas, donated to the British Museum in 1923. She insisted on keeping her personal live specimens sealed in cellophane bags in her bedroom, *'so I can see what they are doing and so that the children do not annoy them'*. As Rothschild explained in the foreword to her classic study on parasites, *Fleas, Flukes & Cuckoos*, these minute jumping insects have had an intimate relationship with mankind over thousands of years. Their impact scales from the simple bed bug that hides in cracks and crevices at night, and *'steals out to suck blood surreptitiously from a sleeping beauty's breast'* to being the prime vector of devastating infections that have killed hundreds of millions from pandemic plague.

Fleas are small parasitic insects that live by sucking blood from their animal hosts. Their ancestors diverged from flies about 160 million years ago, giving up the use of wings, preferring instead to leap from meal to meal using a pair of extremely powerful hind legs. Indeed, for their body size, fleas are among nature's most powerful jumpers, far out-leaping the best human Olympiads. They can cover a distance vertically and horizontally more than 200 times their body length.

Almost inevitably highly mobile fleas that feed on other animals' blood became prime targets for less mobile life-forms that wished to hitch a ride. Fleas, like many flying parasites (in particular, mosquitoes and tsetse flies) carry a wide variety of passengers including worms and bacteria. One in particular, a meddlesome variety called *Yersinia pestis*, has proved the most lethal to humankind. *Yersinia pestis* infects rodents – rats and marmots in particular. Fleas that bite these creatures pick up the bacteria, which then thrive in their guts (apparently without harm to the fleas), only to be spread via flea bites to other mammal hosts, including humans.

The first human pandemic caused by this bacterium, bubonic plague, occurred in the sixth century AD just as Emperor Justinian I was trying to reunite the ancient Roman empire after its disintegration at the hands of Germanic invaders such as the Vandals, Huns and Goths. The outbreak of plague is thought to have originated in Ethiopia, from where the imperial capital Constantinople imported huge quantities of grain. Plague thrived in the narrow streets of the city, where fleas could easily leap off the backs of infected rats on to humans who lived in cramped, grimy, urban conditions. At its height, the pandemic is thought to have killed as many as 10,000 of the city's inhabitants *every day*.

Even this devastation didn't have as great an impact as the second outbreak – the most famous occurrence of plague in all history.

What on Earth Evolved? ... in Brief

The Black Death (1347–51) originated in China, and was spread to Europe by Mongol invaders. Little did Jani Beg, leader of the Mongol horde that attacked the Crimean port of Kaffa in 1346, know what his flirtation with biological warfare would unleash on the hapless continent of Europe. Carcasses of dead Mongol victims of the plague were flung over the city walls on giant slingshots called trebuchets, infecting the Italian soldiers inside. After escaping back to port in Genoa, 90 per cent of the Italian troops died, leaving just enough survivors to pass the disease to the European mainland, where, thanks to the leaping fleas and rampaging rats, it ravaged human populations. As many as sixty million people died in China, forty million in Europe and ten million in Africa.

Huge numbers of books have been written about the impact of the Black Death on world history, coming as it did just prior to the overseas expansion of the European powers. It is beyond doubt that the decimation of populations in countries such as England, France and Italy caused a significant transfer of power away from landowners to the surviving peasants, who could command higher wages for their services. Some were able to free themselves from serfdom. In less populous countries that were less affected by plague, particularly those in eastern Europe, serfdom continued unabated until well into the nineteenth century, partly accounting for the divergent political histories of the two halves of that continent.

Others have speculated that the origins of the Christian Reformation can be found in the lax standards of Church discipline that followed the decimation of clergy during the Black Death; or that new forms of pronunciation in languages such as Dutch and English can be attributed to the mass migration of peasants from the countryside into the towns in their quest for new employment. One theory even proposes that it was by only the tiniest margins that the world avoided being plunged into another ice age because of a significant decrease in carbon dioxide levels and deforestation that followed the annihilation of so many people.

On Biodiversity

There is a small caveat, though, to the trilogy of fleas, rats and bacteria being to blame for triggering everything from religious schism to climate change across medieval Asia, Europe and North Africa. At least two rival theories have recently been proposed that question traditional explanations for the cause of such depopulation by flea-borne disease. One suggests that mass death was the fault of a virus, such as ebola, while another proposes anthrax (see page 23). The once cosy consensus that disease spread by minute fleas was to blame for this most turbulent episode in human history is now a cauldron of hot historical debate.

There is no argument about the third pandemic, though – one that spread through China beginning in 1855, killing more than twelve million people across Asia. Unmistakable evidence of the plague bactcrium *Yersinia pestis* has been traced to the bodies of its victims.

Tsetse Fly

> **FAMILY: GLOSSINIDAE**
> **SPECIES: GLOSSINA MORISTANS**
> **RANK: 37**

Bloodsuckers that helped preserve the last remnants
of human hunter-gatherer lifestyles in Africa

OVER THE LAST 10,000 years, the relationship between man
and nature in Africa has followed a different trajectory from that
of almost any other place on Earth, partly thanks to the exploits of
a specific flying insect. Tsetse are large flat flies that are resistant to
being crushed when slapped, an adaptation most probably provoked
by the relentless swatting of animal tails. Like the mosquito (see
page 130), these flies act as vectors for parasites. Unlike mosquitoes,
though, males and females both enjoy a good bloodsucking feast.
The disease they carry is one that has directly affected the ability of
people to establish settled civilizations, contributing to the reason
why today Africa is one of the least 'developed' parts of the world.

Tsetse are carriers of a single-celled parasite called *Trypanosoma
brucei* that causes trypanosomiasis – a serious illness known as sleeping sickness in humans and nagana in cattle, horses, camels and
pigs. The parasite is spread by these flies as they suck blood from
one infected mammal and pass it on to another with subsequent
bites. Sleeping sickness is almost always fatal unless treated. The
parasite travels to the vertebrate lymphatic system and then crosses
into the nervous system, eventually reaching the brain, where it
induces extreme lethargy and finally death. In cattle, horses and pigs
the parasite impairs growth, milk production and eventually, as in

humans, leads to their sickly, premature demise. Although the tsetse evolved alongside mammals between sixty-five million and fifty-five million years ago, their biggest human impact seems to have occurred in Africa over the last few thousand years.

From *c.* 8000 BC humans in some parts of the world began to experiment with breeding wild animals in an attempt to turn them into compliant domesticated species that would yield an ongoing source of meat and milk (see On Agriculture, page 201). From about 3000 BC powerful new civilizations arose in countries such as Egypt and Nubia (modern-day Sudan) based on the growing of crops and the domestication of farm animals. But due to endemic infection by parasites carried in the tsetse, the practice of living in towns and cities supported by surrounding agriculture and farming bypassed African people living south of the Sahara Desert for thousands of years. Instead, lacking domesticated farm animals, most African people clung to a hunter-gatherer lifestyle until well after the rise and fall of the Greek and Roman empires.

Domesticated cattle introduced from the Middle East to west, central and southern Africa had no natural immunity to nagana. It was only thousands of years later when west African Bantu herders finally succeeded in breeding cattle that could resist the parasites carried by the tsetse that the populations of farming people began to increase significantly. With domesticated herds came copious supplies of meat and milk, supporting ever-larger families and enabling herding nomads to fan out across the continent. By about 500 BC, they had reached the east African coast, where they mastered the art of fashioning tools of iron, further helping to establish their culture as the dominant force in Africa. With their tsetse-immune animals, Bantu people then expanded southwards for more than 2,000 miles reaching as far as the Natal in south-east Africa by AD 300. Eventually, with the addition of crop farming, they created powerful kingdoms such as Great Zimbabwe, built around the

Zambezi river between AD 1200 and 1500. In the process, many of the less populous hunter-gatherer people such as the central African Pygmies and southern African Khoisans were either driven out, enslaved or killed.

Infection by parasites also hindered attempts by medieval and early modern North African, Muslim and European invaders to advance into the interior of Africa. War horses, for many years the world's most potent military force (see page 246), could not survive in the tsetse-infected heartlands of central Africa, where imported livestock had little or no immunity against parasites. Not until Vasco da Gama and other Portuguese explorers ventured by boat around the Cape of Good Hope in the late fifteenth century did European gunpowder introduce a new, even deadlier ingredient. It was only from about the 1870s onwards – a period known as the 'Scramble for Africa' – when steam power on sea and over land finally opened up the interior of this continent, that European invaders were liberated from the need to travel or wage war using easily infected animals. Africa's story since is a relentlessly grim tale of colonization, exploitation, armament and division from which its people still reel today.

Modern humans have waged war with partial success against the tsetse, reducing its numbers through campaigns of eradication using pesticides (such as DDT) or by irradiating males with gamma rays causing them to become infertile, thus reducing populations. Success is manifest in the gradual decline in the reported numbers of people suffering from sleeping sickness (estimated at $c.$ 60,000 Africans). Epidemics still break out sporadically, such as one in Uganda that began in 2008. This latest bout is nothing on the scale of an outbreak between 1896 and 1906 that killed more than one million people.

It is impossible to know how African history would have turned out were it not for this parasite-carrying bloodsucking fly. It is a curious twist of fate to think that without it the last remnants of the

On Biodiversity

Kalahari bushmen's ancient way of life would probably have been swept away thousands of years ago by opportunistic agriculturalists. Without the tsetse, the fate of African people today would certainly be very different indeed.

Oak

> **FAMILY: FAGACEAE**
> **SPECIES: QUERCUS ALBA**
> **RANK: 21**

*A highly robust flowering tree that hosts its
own rich ecosystem*

FEW SPECIES IN the plant kingdom have had as big an impact on the history of humanity as the oak. Reaching far back to the beginnings of recorded history, these flowering trees have captured people's imagination for many different reasons including religion, culture and artifice. More specifically, these are trees that have dominated the minds of people living in the northern hemisphere – in North America and Europe in particular. The reason goes right back to the evolutionary origins of the family Fagaceae (beeches, chestnuts and oaks), which begins about one hundred million years ago.

No one knows precisely where the ancestors of oak trees first emerged. As the super-continent Pangaea started to break up 250 million years ago, it split into two enormous terrestrial crusts. Gondwana was the southern mass, which later fragmented into South America, Africa, India, Australia and Antarctica. Laurasia, the northern half, split into Europe, Asia, Greenland and North America. It was somewhere on this northern mass that oaks evolved, becoming key flowering species that rafted across the northern hemisphere as the continents we are familiar with today gradually spread far and wide.

Oak trees are the terrestrial equivalent of a tropical coral reef (see page 52). A typical oak acts as host to at least thirty species of birds, forty-five types of bugs, and more than 200 species of moths. Insects are especially fond of laying their eggs inside its leaves, which swell up with small growths called galls that provide larvae with nutrition and protection. Tiny parasitic wasps are particularly keen on these breeding grounds, as are the larvae of midges, moths, worms and caterpillars.

Not only is the reproductive strategy used by oaks hugely successful, it is also one of the most generous in nature. A mature oak may yield a crop of as many as 100,000 acorns (a type of seed/nut) a year. The result is a highly nutritious feast for scores of different animals ranging from pigs, deer and bears to squirrels, rodents and birds. For thousands of years wood pigeons have devoured up to 120 acorns each per day; birds have migrated across continents just to join the feast. Magpies and jays enjoy them too, and also profit by dining on the eggs of birds that nest in oaks. Owls and bats make their homes in cracks and holes that open up in the bark of old oak trees, which can survive for as long as a thousand years.

There is more subtlety to the oak's reproductive strategy than simply swamping the ground with a surfeit of nuts, hoping that some take root. Acorns are rich in fats and protein and so animals carry them off to hide away in makeshift food stores. It is important that these animal transporters do not eat *all* the acorns, since the seeds do not pass through the gut undigested. Oaks therefore lace their nutrient-rich acorns with a carefully measured dose of tannin, a chemical that can upset the digestive system when consumed in large quantities. This encourages the animals to bury their cache as a long-term supply of food, and not eat too many all at once. Some animals die prematurely and their acorns remain uneaten. Others forget exactly where their food stores are buried. In this way at least some of the oak's crop of acorns are successfully planted by animals, usually far away from the parent tree, and left to germinate in peace.

Tannin has had its own impact on human history. Vats and barrels made of tannin-rich oak are largely responsible for giving wine and spirits their intense aromatic flavour. Similarly, tannin obtained by soaking wood from oaks in water has been used for thousands of years for preserving leather hides that are subsequently turned into everything from boots to book covers. But the impact of the oak tree on human culture extends a great deal further. These trees, sometimes dubbed 'Lords of the Forest', have been at the core of the mythology and folklore of the people of the northern hemisphere since the beginning of recorded history.

Dodona is the most ancient oracle in Greece, far older than its more famous rival at Delphi. As long as 4,000 years ago, Neolithic farming people gathered there to consult their mother goddess. She spoke to them through the rustling of leaves in the oak forest all around. Herodotus, the Greek storywriter, said that a mighty beam of oak from this forest was used in the construction of Jason's famous boat *The Argonaut*, giving its crew the prophetic foresight they desperately needed to help them complete their epic voyage.

To the Norse people the oak symbolized Thor, the god of thunder, because it was thought to be struck by lightning more often than other trees. One of the most famous Norse outdoor shrines was an ancient tree, 'Thor's oak', which stood proud for hundreds of years. It was worshipped by flocks of Germanic pagans who came to Geismar, in northern Hesse, to seek their own protection against the fickle gods of nature. The tree survived until the year 723 when, according to legend, a passing English Christian evangelist called Winfrid hacked it down to prove that Christ was the one and only true God. When the locals saw that Winfrid was not struck down by lightning for his mischievous deed it is said they converted to Christianity in droves. In such a way did this new religion come to replace ancient Norse veneration for these majestic trees and later Winfrid became Germany's patron, Saint Boniface.

Medieval Christendom's zeal for destroying trees put the relationship between man and oak on to an entirely different footing. From *c.* AD 900 onwards the greatest landowners were the monastic orders, such as the Cistercians, who have been described as the 'shock-troops' of deforestation during the medieval period. Clearing forests to create fields for farming that could then be offered for rent was one of the most successful routes to riches in medieval Europe. Oaks also provided perfect timber for building monastic settlements. Like Norse shrines, Christ's temples were surrounded by oak. Unlike pagan centres of worship, though, they were generally made out of dead wood, and ceremonies took place indoors, not in the living forest.

By the sixteenth century, European axes were falling on the continent's once prolific oak forests with renewed vigour. Wood from oaks was the material of choice for building the ships that took European explorers anywhere and everywhere in the world. The strongly preservative qualities of oak proved ideal for making vessels that were sturdy enough to weather oceanic storms and tough enough not to rot. Britain was once smothered by oak forests from shore to shore. But oak became such an essential part of the new way of life that by the seventeenth century the country was facing a severe deficit. Only the discovery of coal as an alternative fuel saved Britain from a chronic energy shortage, and other timbers from the pine forests of the Baltic and America were sourced to make up for a dearth of suitable indigenous building timber.

There is a final curiosity – a kind of postscript to the story of the highly successful oak. There are about 400 described species of oak, but it is almost impossible to identify one single species as more or less significant than another. The reason is that these are trees that defy the traditional scientific definition of what a species actually *is*. Oaks are among the 10 per cent of flowering plants that have stuck with wind power as their chosen medium for pollination. Compared to the targeted approach of pollination by insects, wind power is extremely imprecise and haphazard. Oaks have overcome this potential negative by defying one of the basic rules in the human understanding of nature. They habitually breed *between* different species, creating perfectly fertile hybrids. If pollen from their own kind doesn't blow their way, pollen from a different species will usually do.

Such promiscuity is another reason for the oak's great success. But it's something that leaves scientists rather scratching their heads wondering: what's the use of a term such as *species* in a world of hybrid oaks?

Acacia

FAMILY: FABACEAE
SPECIES: ACACIA TORTILLIS
RANK: 36

*An angiosperm with superlative powers
of survival and symbiosis*

A TREE STOOD all alone in the African desert for hundreds of years, at least 400 kilometres away from its nearest rival. With the ever-increasing heat and dryness of the Sahara Desert, what was once a rich forest finally faded into history; all, that is, except for a single surviving specimen – the *Arbre du Ténéré* – the most isolated tree in the world. Acacias, of which there are about 1,300 species, are among the most robust flowering plants ever to have evolved. Their range of survival strategies is so impressive that these trees will continue to populate the world despite climate change or human endeavour.

They belong to a legume family called Fabaceae. Like all legumes they can fix atmospheric nitrogen thanks to their symbiotic relationship with *Rhizobium* bacteria (see page 30). Such self-nourishment puts them high up nature's list of die-hards. It also means that their seeds are highly nutritious to eat. These plants do not team up with bacteria alone to maximize their own chances of survival. Several acacia species, especially those found in the dry deserts of Africa, Texas and Central America, play host to vast colonies of ants (see page 161). A queen ant lays her eggs in an acacia's large spiky thorns, which are designed to protect against herbivores. When her workers hatch they act as the acacia's police force, protecting the tree against other insects that might happen to try their luck by feeding on the legume's nutritious leaves or stem. In return for their guardianship, the acacia produces nectar from glands conveniently located to feed the ants. It also secretes small orange highly nutritious beads from the tips of its tiny leaves as additional food for ant larvae.

This is a powerful partnership. Experiments have shown that cutting off an acacia's thorns and removing the ants can cause the plant to wither and sometimes die. The ants will even patrol the perimeter of the plant, destroying the seedlings of any other plants germinating nearby to preserve the acacia's precious supplies of water. Branches of trees that approach from above also get short shrift. Armies of ants march across and devour the rival branch until it is no longer a threat. Devotion, loyalty and fair exchange are hallmarks of this natural association between two unrelated species whose mutual dependence is essential to the survival of each. It is easy in this case to see why and how evolutionary adaptations in nature have occurred, as the bonds of association between such species became strengthened by their mutual success, one generation after the next.

Acacias are pioneering species – a term that applies to life-forms that are able to colonize new lands with the minimum of fuss. Their seeds will happily germinate in the driest, thinnest, most under-nourished soils in the world. They dominate the arid Australian outback and the deserts of Africa and the Americas. If conditions are simply too tough, acacia seeds have been known to lie dormant for up to sixty years, sprouting into seedlings when conditions for life are sufficient to support new growth.

The impact of acacia trees, and the umbrella thorn in particular, has been no less profound in human culture. Wood from this tree is recorded in the Jewish Torah (Christian Old Testament) as being the material that God instructed his chosen people, the Israelites, to use to construct their tabernacle – the Ark of the Covenant. Inside this chest they are said to have placed the Ten Commandments carved by God on tablets of stone that their celebrated leader Moses brought down from Mount Sinai. Later in the Bible an acacia tree bursts into flames – the famous burning bush, sent as a sign from God to Moses.

African people have traditionally used acacias for their every-day needs from making wagon wheels, fence posts and cages, to constructing a type of string from its bark. Its nutritious pods and leaves provide fodder for animals (camels and giraffes particularly) while its sap, an edible form of gum, is now used as a stabilizer in modern, mass-produced foods (additive number E414).

The most famous acacia is still the *Arbre du Ténéré*, the tree that stood for so many hundreds of years all alone in the desert. Its trunk acted as a beacon for travellers for miles around while its roots were a testament to an unsurpassed thirst for survival, stretching down-wards more than *thirty-six metres* to the water table below. But like many of life's great legends, its tragi-comic end eventually came not through its lack of resilience against nature, but through the negli-gence of man, in the form of a drunken Libyan truck driver, who carelessly knocked it over in 1973. Its remains now lie in state at the National Museum of Niger and, in its place, local people have erected a commemorative metal post.

Durian

> **FAMILY: MALVACEAE**
> **SPECIES: DURIO ZIBETHINUS**
> **RANK: 89**

Jurassic fruits that lay claim to being the smelliest on earth

SOME WHO HAVE tasted the durian fruit have said that it's like eating *'pig-shit'*, *'turpentine'* and *'onions garnished with a gym sock'* ... Others say this Indonesian delicacy is like eating *'sweet raspberry blancmange in the lavatory'*. Albert Russell Wallace, the nineteenth-century British naturalist who independently came up with his own theory of evolution by natural selection at about the same time as Charles Darwin, was a little less scathing: *'A rich custard highly flavoured with almonds gives the best general idea of it, but there are occasional wafts of flavour that call to mind cream-cheese, onion-sauce, sherry-wine, and other incongruous dishes.'*

Whatever your sense of smell or your taste bud configuration, the durian fruit is certainly one of nature's most spectacular when it comes to advertising its presence. In the tropical forests of Borneo its smell can be detected from as far as half a mile away. This large fleshy spiky-skinned fruit grows up to thirty centimetres long and can weigh up to four kilograms – all of which accounts for its nickname 'king of the fruits'.

Durian trees are typical of flowering plants that evolved alongside herbivorous dinosaurs such as *Diplodocus*, making the most of their insatiable appetite. The black seeds are surrounded by a rich custard-like sauce that makes durians the most sought-after treat for forest animals including orang-utans, mouse deer, sun bears and even tigers. As the animals feast off the fruit, they swallow some of its seeds, which pass through their guts to be deposited far away in a pile of nitrogen-rich dung.

The survival of the durian long after their original dinosaur seed transporters died off 65.5 million years ago can be put down to the huge investment they have made in producing large fruits full of flavour that advertise their presence so spectacularly through such a powerful odour. But the impact of the durian, or at least its ancestors, may be a great deal more significant, at least according to one theory. Edred Corner (1906–96) was a British botanist who specialized in tropical plants and spent much of his life in Singapore. In his view ancestors of today's durian trees shed light on the 'abominable mystery' articulated by Charles Darwin as to the origin of angiosperms, or flowering plants. According to Corner, red fleshy fruits similar to today's durian were descended from cycads, trees that had already mastered the trick of using insects as pollinators (see page 97). Their durian ancestors took the idea a stage further and developed a technique for spreading their seeds that involved packaging them in well-advertised (red and smelly) fruits that would prove irresistible to passing animals.

Corner's 'Durian Theory', first published in 1949, was an attempt to associate the origins of all flowering plants with the durian lineage. Plants that evolved fleshy edible fruits were therefore the pioneers of the amazingly successful rise of the angiosperms. Corner's theory has been extensively argued, debated, rejected, supported and discussed since he first put it forward. Today it is universally agreed that flowering plants originated in tropical forests, the durian's natural habitat. Whether or not durian ancestors are the evolutionary forefathers of the angiosperms is still open to debate, and other plants, such as the ancestors of today's water lilies, pepper plants and magnolias have more recently been proposed as alternative candidates.

The group of angiosperms to which today's durian belongs is certainly among the most significant of all flowering plants. The rose-like eudicots (or rosids) are a giant group of genetically coherent plants that include many of the world's most valuable flowering trees such as rubber trees, poplars, willows, wisteria, acacias, apples, plums, elms, figs, oaks, beeches, birches, hazelnuts, walnuts, myrtles, eucalyptus, frankincense, myrrh, oranges, lemons, maples and mahogany.

If the durian really was the ancestor of all that lot then it would certainly take pole position among nature's most influential species. And even if its evolutionary credentials are uncertain, this most smelly species still warrants a place in the table.

Bamboo

```
FAMILY: POACEA
SPECIES: MELOCANNA BAMBUSOIDES
RANK: 40
```

*Ancestral grasses that developed impressive
defences against dinosaurs*

STUDYING STOOLS OF FOSSILIZED dinosaur dung
may not sound like one of the most appetizing pursuits in modern
science, but for two American palaeontologists such patient analysis
has recently begun to throw new light on the origin of one of the
most important and widespread group of flowering plants on Earth
– the grasses.

What Dolores Piperno and Hans-Dieter Suess found as they
peered through their microscopes at dung from dinosaurs were the
microscopic remains of an ancient flowering plant defence system
pioneered by some plants to help them avoid being eaten by the
herbivores that roamed the land of what is now India some sixty-
five million to seventy-one million years ago. The system worked by
liberally lacing their cells with fragments of silica (like sand), which
interferes with the chewing and grinding process in a typical herbivorous dinosaur's mouth. Over time, of course, herbivores evolved
ways of circumventing such defences, leading to the various types of
teeth, such as molars, that evolved in mammals once the era of the
grazing dinosaurs came to an end 65.5 million years ago.

Silica-laced stems and leaves are characteristic of monocot grasses
that, until the findings of Piperno and Suess, were thought to have
evolved only *after* the demise of dinosaurs. The earliest complete

fossil evidence of grasses dates back only as far as fifty-six million years. However, thanks to the discovery of those tell-tale silica fragments, called phytoliths, in dinosaur dung, it is now thought that grasses emerged considerably earlier than previously assumed, probably during the late Cretaceous Period.

Further phytolith study reveals that most primitive grasses were ancestors of today's bamboos, among the fastest-growing plants in the world. Today a well-watered bamboo can put on a good fifteen centimetres of growth a day and reach up to forty metres in height, although one living species has been known to grow by more than a metre in a twenty-four-hour period. Back in the days of the dinosaurs, some species grew even faster – perhaps as much as several metres a day, with an ultimate height of more than sixty metres. Such prodigious growth rates were attempts to survive being grazed to the ground by greedy dinosaurs. Grasses that prospered best were those that grew the fastest. Bamboos, like all grasses, also thrived thanks to the monocot design for defence in which their all-important growth buds were tucked away safely underground. Regardless of how savagely grasses like bamboos were grazed, their newest growth was always least vulnerable to attack.

These survival characteristics – silica, rapid growth and a submerged growth bud – are the hallmarks of what have made grasses great. As the climate has dried out over the last forty million years more than 30 per cent of the world's landmass has come to be covered in different species of grasses, many of which have become essential foodstuffs for humans following the start of farming 12,000 years ago (see On Agriculture, page 201). Before humans began to cultivate grasses to produce wealth for settled societies, bamboos were among the world's most successful and influential plant species. As many as 1,000 types of bamboo exist today, most in east Asia, Australia, India and Sub-Saharan Africa. Huge forests of fast-growing woody bamboos are thought to have dominated many of these landmasses before the arrival of humans.

On Biodiversity

Although the evolutionary significance of today's bamboo is clear from the analysis of dinosaur dung, it is less easy to know whether other grasses (such as wild wheat, oats and barley) originally evolved from ancient bamboos or other sister groups. However, further evidence in favour of the bamboo's ancient lineage comes from their curious flowering habits. Bamboos flower infrequently – sometimes only once every sixty to 130 years. When they do bloom *they all flower together* sometimes regardless of geographical location. No one quite knows how they achieve this – do they possess some kind of mysterious genetic programme that acts like a clock? Are they somehow capable of sending a signal to each other? The strategy probably emerged as a response to the huge grazing pressure applied by dinosaurs seventy million years ago. By producing so many flowers and seeds all at once, some species of bamboo literally over-whelmed the appetites of herbivorous animals, ensuring that at least some of their progeny had a good chance of survival into the next generation. As a final gesture of goodwill, the parent plants of some species automatically die off after flowering so as not to compete with their fast-growing seedlings for water, light and space. Such is the power in nature of a species' effort to ensure its continued genetic inheritance regardless of any concern for the longevity of an individual's life.

Fruit-filled forests with tall woody trees that suddenly die off cause profound ecological dislocation. The effect is felt acutely in places such as the Bay of Bengal, which suffers huge surges in its rodent population whenever its bamboo forests flower, fruit and then die every thirty-five years. Vastly increased rodent populations munch their way through other crops, leading to famine. Parasites that live on rats (see page 180) cause a major upturn in the spread of diseases such as plague and typhus that are often fatal to humans.

The occasional mass suicide of bamboo forests would not matter nearly so much were it not for the fact that these fast-growing grasses are of such great social and economic importance. In some Asian societies bamboos have completely transformed human ways of life. They are used for everything from cooking (bamboo shoots) and medicine (as a natural antiseptic) to construction (roofs, flooring, scaffolding, gutters and water channels). Thanks to the tensile strength of bamboo, Chinese engineers operating over 1,500 years ago were even able to make cables for lowering drill bits into rigs for mining natural gas. Then there's the plethora of other traditional uses for bamboo that includes everything from making fences, bridges, canoes, walking sticks and furniture to musical instruments, hats, weapons, fishing rods, baskets and knitting needles.

Honey Bee

FAMILY: APIDAE
SPECIES: APIS MELLIFERA
RANK: 33

Industrious transporters responsible for pollinating many of the world's flowers and for producing pots of honey

TODAY'S WORLD WOULD look very different were it not for the wonder of honey bees. As many as 20,000 different species have been recorded, but it is *Apis mellifera*, the European honey bee, that has had the greatest effect, not just on human history but also on many of nature's most prolific flowering plants.

Honey bees evolved more than one hundred million years ago – in the days when ten-metre-long dinosaurs dominated the landscape, pterodactyls the size of small aeroplanes swooped down from the skies and the seas were filled with exotic ammonites and terrifying ichthyosaurs. The bees' long evolutionary history was recently revealed by the discovery of a one-hundred-million-year-old amber fossil found deep inside a mine in northern Burma. Bees are descended from wasps and, like them, developed some of nature's first social gatherings, displaying many characteristics also seen in human civilizations, which first formed a mere 6,000 years ago.

A honey bee nest contains three types of bee. First is the queen, who lays as many as 500,000 eggs in her three-year lifespan. Most of these hatch out to become female worker bees, which share a range of tasks such as cleaning, feeding, nursing, finding food, storing food and making the wax cells that form the honeycomb. A few, called drones, are males and have just one function – to impregnate the queen, usually in an elaborate ritual that takes place hundreds of metres up in the air. The ritual is so physically demanding that only the strongest males can successfully mate. So much effort is required in this process of insemination that drones usually die soon after sex. This is a woman's world.

The queen communicates with her workers by secreting pheromones that are passed around the nest. Individual bees communicate through the language of dance. The 'round dance' means that food is within fifty metres of the hive. The 'waggle dance', which may be vertical or horizontal, provides specific details about the distance and direction of a particular food source. The 'jerky dance' is used to discuss whether to increase or decrease the amount of food gathering the bees need to do, depending on the hive's overall requirements.

In fact, honey bees are so highly developed that when they need to decide where to locate a new colony they produce the earliest known example of what in the modern human world we might call a 'democracy'. Various scouts will identify potential nesting sites and then report back the location of each one to the rest of the community by means of their various dances. Other bees check them out, before returning to the nest and dancing longer and harder in the direction of the site they think is the best. After about two weeks, the site with the strongest, most vigorous dance is the winner and the colony swarms. It has been estimated that using this form of range voting, the bees will choose the best location approximately 90 per cent of the time.

Other traits familiar in human societies even stretch to ills such as alcoholism. Bees that imbibe fermented nectar are notorious drunkards whose behaviour is similar to that of an intoxicated human. Indeed, their antics mirror the effect of alcohol on the human brain so closely that scientists now study intoxicated bees in their efforts to research the effects of alcohol abuse in mankind. Guard bees stationed outside a hive prevent such drunkards from causing trouble and – like bouncers – banish them until they have sobered up.

Repeat offenders are not tolerated – their legs are chewed off as a punishment.

The co-evolution of flowers and bees during the Cretaceous Period is one of nature's best examples of evolutionary teamwork. Without the targeted pollination services of bees, many flowering plants could never have spread themselves so widely across the world. Without these flowers, bees would never have evolved away from parasitic or predatory wasps to feed on a vegetarian diet of nectar and pollen instead.

Humans have harvested honey for thousands of years, well before the earliest civilizations, especially in Africa, where hunter-gatherer people followed the paths of honey-diviner birds that led to nests full of bees. Sealed honey pots have been found in the tombs of pharaohs and the art of beekeeping was a favourite topic among Greek and Roman writers. The oldest artificial hives found by archaeologists date back to *c.* 900 BC in Israel during biblical times – evidence of the Promised Land of 'milk and honey'.

Bees contribute billions of pounds to today's global economy. Their value is not limited to honey production (China is today's largest producer, harvesting more than 300,000 tonnes a year) because bees are also responsible for pollinating so many of the most common and popular fruits grown for human consumption. The practice of travelling with hives of bees from one US state to another and hiring out swarms for pollination is big business. Migratory beekeeping is estimated to support some fifteen billion dollars of fruit crops in the United States alone.

There is, however, a sinister sting in the tail to the story of today's European honey bees. Since 2006 colonies have been mysteriously disappearing, reducing by as much as 50 per cent in some areas of Europe and America. If such Colony Collapse Disorder (CCD) continues, human farmers will face incalculable losses in terms of the free pollination services traditionally provided by these ancient flying insects. Concerned Western governments are now pumping millions into scientific research in an effort to discover its cause. (Is it the overuse of pesticides? Increasing urbanization? A mysterious viral outbreak?) If the agricultural lives of humans are not to be made a great deal more complicated, experts need to find a remedy – and fast.

Ant

```
FAMILY: FORMICIDAE
SPECIES: ATTA SEXDENS
RANK: 25
```

Engineering geniuses that have successfully adapted
to just about every habitat on Earth

WHEN THAT METEORITE struck the Earth 65.5 million
years ago, devastating the dinosaurs, pterosaurs and so many other
species that dominated the land, nature's biodiversity was its best
insurance against total annihilation. Certain genetic combinations,
manifested in different types of organisms with varying ways of life,
found that despite the devastation all around, life went on pretty
much as usual. It is common knowledge that mammals thrived
in the absence of large reptilious predators, but less well known is
that several other animal families also came into their own after
the demise of the dinosaurs. Although far smaller than mammalian
species, some of these creatures have ultimately been just as significant.
In particular, insects called ants.

Ants dominate life in the soil on every continent except frozen
Antarctica. Only Iceland, Greenland and a few Pacific islands have
no indigenous species. In many ecosystems, especially in the tropics, up to 20 per cent of the mass of all land-living creatures can be
accounted for by ants alone, putting them at the top of today's list of
nature's most populous life-forms.

Fossils of ants trapped in amber date back to the mid-Cretaceous
Period (*c.* 120 million years ago), but it was only after the demise of
the dinosaurs that ants spread in significant numbers. About 14,000

different species have been identified to date. Having learned to live in large societies so long ago, sufficient time has elapsed for these creatures to have explored every conceivable form of successful society. Therefore, there are few better natural parallels to human social instincts than observing the various practices of ants.

At the centre of most ant societies is a queen who, soon after she reaches maturity, indulges in a once-in-a-lifetime nuptial flight in which she mates with many males, building up a store of sperm that she uses for the rest of her life (up to thirty years in some cases) to produce fertile eggs. Once the male ants have flown and mated (or attempted to) they quickly die off. The queen then burrows into the ground and begins to lay her eggs. Most of them are fertilized and produce females which will become either future queens or sterile workers. Other eggs she leaves unfertilized, and these then grow into male clones.

Some experts believe sterility in female workers evolved because by caring for their queen (and their sisters) the chances of their genes surviving into future generations are statistically greater than if they had offspring themselves. Whatever its origins, sterility seems to act as a catalyst for selfless loyalty, co-operation and dedication – key components of an effective society.

Farming crops and animals is a habit usually thought to be exclusive to people – a way of life that began about 12,000 years ago and is often believed to set mankind apart from the rest of nature. That is not the case. Ancestors of ants that belong to the tribe *Attini*, which live mostly in South America, formed nature's original and most prolific farming communities. For many years experts believed that these leafcutter ants simply foraged for plants they preferred to eat, or perhaps used the leaves to protect their nests from rain. But thanks to the observations of a contemporary of Darwin's, British geologist Thomas Belt (1832–78), it was discovered that these ants forage for leaves to use as manure for cultivating a certain type of fungus (*Rozites gongylophora*) in neatly kept gardens buried deep inside their nests. These ants not only feed on this fungus but they also regularly pluck tufts of its mycelia to plant in new areas of their gardens, where they grow rapidly, continuing to provide the nest with a rich source of food.

So vital is the practice of cultivating food for these pioneering agriculturalists that before a new queen embarks on her nuptial flight she tucks a store of mycelia into her body for safekeeping. When she begins the process of building a new nest, her priority is to retrieve her wad of fungi, which she uses to establish her colony's first garden – even using her own excrement as a form of manure to stimulate its growth.

Ants that cultivate crops also learned how to domesticate animals, like humans have done many millions of years later. Aphids are to ants what cows are to humans. Being specialists in sucking out the sap of plants, aphids excrete a highly nutritious juice that ants 'milk' from them by stroking the aphids' back legs with their feelers. In return for providing a source of milk, ants protect the aphids from enemies (just as humans protect cows by fencing them in or keeping them in barns). So close has the association between aphids and ants become that some aphid species have even grown to mimic the face of an ant, to further facilitate the feeding process. Other symbiotic associations have arisen between ants and flowering trees such as acacias (see page 147).

If farming and agriculture are social practices that evolved in nature millions of years ago, then so too was the insect equivalent of supporting a society through nomadic expansion and war. In the thirteenth and fourteenth centuries Mongol chief Genghis Khan and his blood relations established the largest contiguous empire the world has ever known, stretching all the way from China to the Mediterranean. Their tactics of moving in hordes from place to place, subsuming other civilizations in their path, was a way of life that originated in microcosm among types of ants, such as *Eciton burchelli*, that thrive in Brazil, Peru and Mexico.

The nests of these 'army ants' are, like those of human Mongols, a marvel of self-sufficiency. Thousands of ants link up like threads to form a chamber called a 'bivouac' in which the rest of the colony lives. When it is ready to feed, the nest breaks up, almost in an instant, and recasts itself into war-like hordes that swarm over areas of the forest floor capturing, paralysing and devouring all animal life in their path. Other insects, snakes, even birds and monkeys can be trapped, killed and eaten by these huge armies of marching, munching ants.

As in any natural self-organizing system, there are no leaders. Individuals take turns at the front of the swarm, then move back, allowing others to take their place. Group co-ordination and problem-solving is achieved through the passing of pheromones – chemical smells – which operate as a signalling language, spreading instructions from one individual to another. Like a slime mould (see page 36), these individuals are so co-ordinated that they act as if constituent parts of a super-organism, all the stronger for never having to depend on centralized control, and highly efficient at solving the puzzle of where to find the most food. Feelers are so much more powerful when they act in a concert of literally millions of co-operating pairs.

Finally, there are ant societies whose livings depend entirely on conquest. The Roman empire and its daughter project, European overseas colonization, provide reasonable human equivalents. Legionnaire ants (*Polyergus breviceps*) are slave-raiders incapable of growing or even feeding themselves. Like the most affluent in Roman society, they have their food served to them by slaves. They establish their colonies either by stealing larvae from rival nests or through tactics reminiscent of a classic human *coup d'état*. A queen will sneak inside another nest, kill its ruling incumbent and take over, instructing her new subjects to bring up her own offspring instead.

The impact of ants on the evolution of other species they come into contact with has been profound – be it symbiotic, parasitic or predatory. The societies they have succeeded in building are the oldest, most robust and versatile in all nature – far out-performing in numbers and efficacy those experimental equivalents built by humans in the last few thousands of years. With supreme success at surviving mass extinctions already proven after the meteorite strike 65.5 million years ago, these are creatures whose versatile ways of life and ability to combine as super-organisms are likely to feature near the top of anyone's list of nature's best long-term bets.

8

On the Rise of Reason

How the mental skills of mammals gave rise to a new evolutionary force that turned monkeys into men.

Bat

ORDER: CHIROPTERA
SPECIES: DESMODUS ROTUNDUS
RANK: 75

*Flying musicians that have successfully adapted
to most available habitats*

AUSTIN, THE STATE capital of Texas, has a number of claims to fame. Perhaps the most surprising is the spectacle found underneath Congress Avenue Bridge, where a colony of about 1.5 million bats lives – the largest in the United States. They stay there during the summer and migrate to Mexico for the winter. More than 100,000 tourists come to see the sight each year.

Bats are the only flying mammals. There are gliders, like flying squirrels and colugos, but sustained flight seems to have evolved just once in the mammal world, representing evolution's fourth and most recent sortie making use of the combined forces of wings, muscles and air resistance to achieve aerial mobility (see also dragonfly, page 101; *Quetzalcoatlus*, page 118; and *Archaeopteryx*, page 121).

The earliest bat fossils date back nearly sixty million years, with a major population increase occurring about fifty million years ago during a period of sudden and dramatic climate change when global temperatures soared by as much as six degrees in just 20,000 years (see *Azolla*, page 87). As the world got warmer, insect populations soared, ideal for creatures like bats that feast on them. Fossils show clearly that by this time bats had already learned how to fly. With such an abundance of insect meals to devour and their consummate skill in 'seeing' in the dark using

On the Rise of Reason

high-frequency sonar, bats developed into some of the most successful creatures on Earth.

Bats inhabit every part of the world except the polar regions. More than 1,000 types are known to science today, representing nearly 20 per cent of all mammal species, the second most diverse group after the rodents (see page 180). Some are as small as a bumble bee, others have wingspans that reach more than 1.5 metres. The most ancient lineage, microbats, mainly eat insects, while the more recently evolved megabats mostly feast on fruits such as figs and play a vital part in many tropical ecosystems by pollinating plants and spreading seeds.

Humans don't tend to think of bats as musical creatures because most of the sounds they make lie far beyond the scope of human hearing (between 14,000 and 100,000 hertz – people can hear only up to 20,000 hertz). Some bats make continuous tones of various notes. In others the tone modulates and can be emitted as a harmonic chord. Bats learned to make music because it helped them hunt in the dark when they were safest from attack by predators. Differently pitched sounds help bats judge the whereabouts and proximity of solid objects, using time delay and pitch change as cues to build up a mental image of the world without the need for light or sight. This technique has a name – echolocation.

A bat's song comes from its larynx. After its ears pick up the echo of its song, its brain calculates the time difference between each ear to work out where an object is located. Differences in pitch between the sound when emitted and the echo also tell the bat if the object is moving or not – very useful when trying to find flying insects in poor light. So skilful are these creatures at 'seeing' by sound that some species are able to distinguish between two separate targets less than half a millimetre apart – all in the pitch dark.

Bat species are endangered today as a result of habitat loss, owing to human urbanization. The use of pesticides on fields also poisons insects that are then eaten by predators such as bats (see also earthworm, page 93). Conservation efforts have been under way since the 1960s, and even include attempts to convert British World War II pill boxes, originally designed to defend against German invasion, into bat shelters.

A more recent hiccough in the human–bat relationship is the growing realization that some bats are, like rats and mosquitoes, vectors of disease. A few bat species have diversified their tastes and become bloodsucking vampire bats (*Desmodus rotundus*, for example). They bite other mammals (such as cows) with a slash from the upper teeth or by nipping off a small piece of skin. The bat then feeds for up to half an hour from the same wound while anti-coagulants in its saliva prevent the blood from clotting.

Rabies is a virus spread by vampire bats. It passes from bat saliva into the blood system of a mammal host (such as a dog or human), causing it to salivate and burst into fits of biting, tactics designed to help spread the virus via saliva. Bats are also thought to be vectors of other diseases such as ebola, a deadly haemorrhaging virus that affects humans and is currently annihilating populations of African gorillas. SARS, the often fatal respiratory influenza-like disease that nearly caused a human pandemic in Asia in 2003, is also thought to be carried by bats, which act as a persistent reservoir species for the virus.

As the only mammals that can fly and with more species than almost any other group of mammals, the extraordinary lifestyles of bats, honed by their musical pitch in the blackness of the night, make these creatures among the world's most remarkable aviators.

Sperm Whale

ORDER: CETACEA
SPECIES: PHYSETER MACROCEPHALUS
RANK: 59

*A deep-sea swimmer that has learned to
survive the onslaughts of man*

'CALL ME ISHMAEL' is one of the most famous first lines
in American literature. Herman Melville's nineteenth-century
pre-Civil-War classic tells the story of a sea captain, Ahab, and his
vendetta against an aggressive male sperm whale called Moby-Dick
that had bitten off his leg. Ishmael, the narrator, is a crew member of
the whaling ship *Pequod*, which sets off in pursuit of the whale but
is eventually sunk by the creature after an epic three-day mid-ocean
duel. Man's attempt to control nature lies at the heart of Melville's
tale and in this story mankind is the loser.

The story behind Moby-Dick was based on fact. In 1820 an
American whaling ship, the *Essex*, was rammed and sunk by a large
sperm whale 2,000 miles off the west coast of South America. Eight
crew members survived. One of them wrote about the episode and
his account gave Melville the context for his classic tale.

Scientific researchers have recently been trawling through the
records and have found that the *Essex*'s fate was not unique. A similar incident occurred in 1851, when another wooden whaling ship,
called the *Ann Alexander*, was rammed and sunk. After the introduction of steel-hulled ships and explosive harpoons in the second half
of the nineteenth century, the scales in the contest between man
and whale tipped in mankind's favour. Since the mid-eighteenth

century whales, among them blue whales, the largest creatures that ever lived, have been subject to severe human predation.

There are two types of whale – those with teeth (the dolphin and sperm whale families, for example) and those without (filter-feeders, such as the baleen and blue whale). Both are derived from a land-based common ancestor that also led to the Artiodactyla, an order of hoofed mammals that includes today's pigs, camels, deer, giraffes, sheep, goats and cattle. Fifty-three-million-year-old fossils of a creature called *Pakicetus*, found in Pakistan, are thought to represent the whale lineage shortly before it returned to the water. Genetic analysis suggests that the hippopotamus is the closest living land relative to the dolphin and whale.

Whales and dolphins are descended from hoofed wolf-like animals that later 're-adapted' to live as swimmers in the sea. For these creatures at least, returning to the seas fifty million years ago must have seemed a safer bet than staying on the land, which was by then becoming overrun with mammals of ever-increasing size and numbers. Large birds of prey came to dominate the skies after the demise of the pterosaurs, perhaps providing another evolutionary pressure on these creatures to try survival by submergence.

Over millions of years the nostrils of these creatures migrated to the top of their heads to become blowholes. Bones that were once their legs, hooves and pelvis became redundant and sank deep into their bodies, helping to streamline them for a life spent swimming. Some whales also evolved a ball of oily fat in their heads to help focus the production and detection of sounds underwater, which in many species served the equivalent function of echolocation in bats (see page 168).

Relatively large brains, with the capacity for using rational thought as a tool for survival, explain why whales (like elephants, see page 175) are among the most successful creatures on Earth. Most whale species are highly social. They communicate either by using a series of clicking sounds or by singing long, repetitive twenty- to thirty-minute songs, mostly voiced by males in the winter months.

Some whales use 'culture' and showmanship to win the favour of females during the mating season. These are the less aggressive, 'toothless' baleen whales, such as the humpbacks, which sing complex, haunting songs. Although no one is 100 per cent sure of the purpose of whale song, most experts believe it forms part of the various rituals used by males to try to win the approval of potential female mates.

In some whale species brute force is what counts most when it comes to making the best impression. The biggest, strongest, most successful males take part in head-to-head combat from which only those that are successful go on to mate with females. Such behaviour explains why, over the course of many generations, male offspring have genes selected for size and strength leading to sexual dimorphism – a condition in which one sex is typically much larger than the other. Male sperm whales are often up to *twice as large* as females.

An adaptation resulting from a male sperm whale's violent mating ritual is its characteristically swollen head, which is filled with a thick, white oil. This feature is now thought to have evolved at least in part as a battering ram to deliver blows and absorb the shock of head-on attack during duels for damsels (focusing sounds for echolocation and providing buoyancy are other uses).

It was an unfortunate by-product of the evolutionary process that this oil, known as spermaceti, proved so attractive to human hunters for use in everything from making candles to lighting lamps. A large sperm whale could yield up to *three tonnes* of spermaceti, once its head had been cut up on a whaling ship. Human hunters would scoop out the oil into barrels before taking it to market onshore. At its height, in the late eighteenth century, the American whaling fleet numbered about 360 ships and employed as many as 5,000 crew, who produced 45,000 barrels of spermaceti annually (compared with 8,500 barrels of conventional whale oil). Spermaceti was also an important lubricant for machines used in cotton mills during the early Industrial Revolution and the main fuel for public lighting in the United States until the 1860s when it was replaced by lard oil and later petroleum.

Sometimes sperm whales fought back – as the crew of the *Essex* found out to their cost. With the deployment of explosive harpoons, however, the contest became one-sided. By 1880 nearly a quarter of the estimated 1.1 million sperm whales worldwide had been killed. Stocks bounced back, however, as petroleum and other products replaced the more expensive and hard-to-acquire spermaceti. But following World War II whaling became popular again and sperm whale numbers plummeted by as much as 33 per cent. It is estimated that at least 700,000 sperm whales were slaughtered for their precious oily heads between 1946 and 1980. Human sentiment began to change only when scientists found that whales were cultured, intelligent creatures that could sing, click and, in their own way, talk, leading to the advent of the worldwide whaling ban that began to take effect in the late 1960s.

Elephant

> **FAMILY: ELEPHANTIDAE**
> **SPECIES: LOXODONTA SP.**
> **RANK: 44**

Lumbering giants with big brains, powerful emotions and a prodigious memory

ELEPHANTS ARE AMONG the strongest, most emotional and highly intelligent creatures on Earth. Their brains are larger than that of any other living land mammal (including humans) with a mass of just over five kilograms. Correspondingly, their capacity for memory, learning, emotions, play, problem-solving and socializing is thought by many to be the most advanced of any non-human species.

Two types of elephant exist today – African and Indian – although many others have lived but become extinct over the last few thousand years. Woolly mammoths, a related species specially adapted to intense cold, died out soon after the last ice age melt as a result, it is thought, of a combination of climate change and over-zealous spear-throwing *Homo sapiens* (see page 191).

Like humans, elephants have a prodigious capacity to learn as they grow up. Most mammals are born with brains that are 90 per cent of their final size. Elephants are born with just 35 per cent of their final capacity, the other 65 per cent developing during the first ten years of life (it is slightly more for the human brain, which puts on 72 per cent). Thus the elephant's ability to modify instinctive behaviour by learning through experience has made it (along with dogs, whales, chimpanzees and, of course, humans) special in nature, giving it the advantage of a more adaptive survival strategy. Rational thought offers the ability to 'learn' during a single generation in order to develop new patterns of non-instinctive behaviour. That's why from an evolutionary point of view, in a rapidly changing world, a big brain can make good survival sense.

Large brains with a capacity for developing off-the-cuff survival skills may also exhibit a capacity for foresight. An elephant can use its trunk as an arm for making wooden tools. Some have been observed digging out waterholes. They then plug these holes with a ball of chewed-up bark and cover them with sand to prevent the water evaporating and so preserve them for future use. The same elephants have later been observed returning to their artificial waterholes at a later date, taking out the plug and helping themselves to a refreshing drink. Captive elephants have even been known to remove their shackles and then make their escape only once they have looked around to ensure no one is watching. Meanwhile, wild elephants communicate over vast distances using sensors in their trunks and feet that pick up signals in the ground sent as vibrations by other more distant elephant groups. These low-frequency infra-sounds travel more effectively through the ground than through the air.

Creatures with large brains usually take years to learn from experience. Like humans, therefore, elephants rely on strong social support systems to care for the young. A mother elephant has to eat huge quantities of vegetation each day to keep her supply of milk flowing, so she needs the support of other adults to help in the tuition and care of her offspring. Strong emotional bonds, indicated by an enlarged region of the elephant brain called the hippocampus, are instrumental in maintaining the integrity of such groups. Devotional behaviour, such as burying the bones of their dead in branches, leaves and sand, has been observed in elephants, as has grief for the loss of loved ones, expressed by standing mournfully over graves and returning to visit ancient burial sites. Humans are the only other species in which such rituals are known, beginning with the Neanderthals (*Homo neanderthalensis*, see page 193).

The group of placental mammals to which elephants belong, Afrotheria, emerged in Africa about 105 million years ago. Along with their extinct ancestors (such as stegodons, mastodons and mammoths), these creatures have had a dramatic impact on the environment of the continents where they have traditionally dwelt – Asia, Africa, Europe (Crete) and the Middle East. As the strongest animals on land, these creatures can literally rip out trees with their trunks. Deforestation by elephantine herbivores creates space in the tree canopy through which new shafts of light can penetrate, providing the all-important conditions in which seedlings can thrive. When modern humans deforest an area, they build on it or farm it, thereby preventing trees from regrowing. Yet, when nature's lumberjacks – elephants – unearth tracts of forest they give it ample time to recover before returning many years later.

Just as brutish elephant strength supports a healthy forest ecosystem, their copious quantities of faeces support other lifestyles on an altogether smaller scale. Large dollops of elephant dung provide ideal foundations for the nests of termites, insects descended from cockroaches. These nests, which can contain upwards of *two million individuals*, make an extraordinary sight, dotted in towers sometimes more than two metres high across the African plains, each one seeded in the wake of a passing elephant's waste.

Elephants have had no less an impact on human history. Ancient Indian, Persian and Phoenician armies were usually fortified by many thousands of elephant cavalry, most popularly by Indian chief Chandragupta Maurya. His 9,000 elephants were, according to Greek historian Plutarch, instrumental in deterring invasion by Greek adventurer Alexander the Great in the fourth century BC. Two hundred years later, Hannibal, the Carthaginian general, deployed a now extinct species of smaller African elephant (*Loxodonta africana pharaohensis*) to cross the Italian Alps in his bid to shatter the Roman empire. So enduring has the military contribution of elephants been that they were still being used by the British army in World War I to haul artillery across munitions yards.

Elephants have traditionally been hunted by humans, not so much for their meat, but for their precious ivory tusks. Elephants use these extended, modified teeth to dig for water, uproot trees or, occasionally, as weapons. For thousands of years ivory has been one of the world's most popular craft materials used to make everything from decorative ornaments and religious artefacts to pipes for smoking opium, piano keys and billiard balls. This easy-to-carve material has been a highly prized commodity for Asian merchants in particular, much of it traded along the ancient silk route that connected markets from Africa and Europe to the Far East.

Objects previously made from ivory are now more often constructed from plastic. Intensive lobbying by animal conservationists led to a worldwide ban on the sale of ivory in 1989, following fears that both African and Indian elephants would soon be hunted to extinction unless action was taken. African elephant populations have since begun to recover. As many as 500,000 elephants are thought to be alive in the wild today, their populations growing at an annual rate of about 4.5 per cent. Human efforts to co-ordinate the worldwide ban on hunting elephants (and whales, see page 174) demonstrate the capacity for creatures with large brains to act 'altruistically'.

Elephants exhibit this trait too. Joyce Poole, an American elephant conservationist, reports how a ranch herder's leg was broken when a female elephant knocked him over with her trunk. On seeing that he was injured, the elephant *gently moved him several metres and popped him under the shade of a tree. There she stood guard over him through the afternoon, through the night and into the next day. Her family left her behind, but she stayed on, occasionally touching him with her trunk.*

Charles Darwin, in his *On the Origin of Species*, stated unequivocally that it was contrary to the laws of natural selection for any creature to behave altruistically, that is '*for the exclusive good . . . of another*'. Many biologists have assumed that selfless acts of charity on behalf of other living things cannot therefore be natural behaviours. Yet the power of rational thought – which operates not over generations but within a single lifetime through that individual's ability to learn – does not necessarily obey this rule. Mammals with larger-than-average brains (such as elephants, whales, dogs, monkeys and humans) do seem to have within them the capacity to override genetic instincts and behave altruistically, be it an elephant protecting a vulnerable human or *Homo sapiens* lobbying for an international ban on the hunting of elephants for ivory.

Rat

ORDER: RODENTIA
SPECIES: RATTUS NORVEGICUS
RANK: 32

Roguish rodents whose intelligence and adaptability have been mostly misunderstood

SEWERS, FILTH, VERMIN and disease are just a few of the many negatives that are conjured up in the minds of most people in the Western world when they think of rats. The historic association goes back to the medieval bubonic plague (see page 136). Rats never featured in the original story of the Pied Piper of Hamelin, who apparently stole 130 German children in 1284. By the sixteenth century, however, their place in the tale was firmly established. By this time these vermin infested every overcrowded street in Europe's burgeoning cities.

Rats belong to the order Rodentia, which comprises more than 40 per cent of all living mammal species and includes mice, squirrels, chipmunks, porcupines, beavers, hamsters, gerbils, guinea-pigs, chinchillas and groundhogs. They are distinguished by their two front incisor teeth, which grow continuously throughout their lives. Therefore survival for these creatures means constantly gnawing through anything they come across – from fully grown trees to plastic-coated electric wires – just to keep their teeth in check.

Rat populations increased with the spread of the world's great grasslands, which in turn arose as a response to the world's ever-cooling climate over the last thirty million years. These are creatures that can eat insects or fruit, but mostly thrive on seeds. So when

On the Rise of Reason

humans indulged in their first experiments with farming crops about 12,000 years ago, and began to store their agricultural wealth in granaries and barns, it was as if rats had won life's lottery. Ever since, the success of rats has mirrored the rise of human civilizations. In Britain there are an estimated eighty-three million rats, twenty-three million more than humans. No one knows the global figure, although some estimates suggest there may be as many as a hundred million in the sewers beneath New York city alone.

Success for rats wasn't just down to hanging off the coat-tails of humanity. Rats are among nature's most prodigious reproducers. Gestation takes just twenty-one days from conception to the birth of a litter that may number as many as ten young. Even more extraordinary is that females are ready to conceive again within just twenty-four hours of giving birth. With a potential output from a single female of more than one hundred offspring per year, numbers can rise dramatically.

There are more than 300 rat species alive today, although two predominate. The black rat (*Rattus rattus*) and the brown rat (*Rattus norvegicus*) both originated in Asia and have spread around the world, often independently of man. Black rats reached Australia thousands of years ago by rafting or swimming, rapidly becoming the predominant placental mammal. Being small, shy creatures that are not fussy about what they eat, rats have frequently been the uninvited guests of human travellers, who during the last few thousand years have transported them all over the world along with their precious supplies of food. Black rats reached Europe by AD 600 accompanied by traders following the silk route. Brown rats arrived later, perhaps by about AD 1600, but have since become the most successful, dominating most habitats except those in tropical regions.

Human–rat relationships vary a great deal from culture to culture. Generally, rats are prized in the East and persecuted in the West. Today many poor people in Asia, especially in Cambodia and

Vietnam, depend on rats as a rich source of protein in their diet. In Ghana rats are both farmed and hunted for their meat. Traditional tribes such as the Irula, of Tamil Nadu in southern India, who work in rice fields, smoke rats out of their burrows not only to protect the rice crop but also so they can cook and eat the animals. Lord Ganesha, a popular Indian deity, is often depicted with a rat at his side. The Karna Mati Temple in northern India, which is dedicated to him, is jam-packed with rats which, it is believed, are destined for re-incarnation in the next life as holy men. Pilgrims are therefore encouraged to eat and drink grain and milk that has been touched by the rats as a blessing.

As rats from the east spread westwards, attitudes towards them hardened thanks to the rise of Judaism and its successful offshoot Christianity, which regarded them as unclean. Touching – let alone eating – rats was strictly taboo in the Judaic culture of the Middle East: '*Any creatures that teem along the ground are vermin: they shall not be eaten*' (Leviticus, chapter 11, verse 42).

By the sixteenth century rats had became associated with diseases such as plague, which is roughly when they appeared as embellishments to the original Pied Piper of Hamelin story. The charismatic musician's primary purpose in the modified tale was not to steal children but to catch rats. In Victorian England rat-catching became big business. The creatures were caught to be sold to London gamblers, who ran as many as seventy rat pits. Punters would take bets on how many rats a bull terrier could kill in one minute. The world record was held by a dog called Jacko, who on 29 August 1862 polished off sixty rats in two minutes forty-two seconds – an average kill time of one rat every 2.7 seconds.

Today, the laboratory rat (*Rattus norvegicus*) is prized by scientists because it is so easy to study, drug and genetically modify for the benefit of research into disease. Such experiments generally cause less public concern than those that involve larger mammals, such as monkeys. Few people in the West care much about what happens to rats, with the exception of those who keep domesticated varieties, also derived from scientific research, as fancy pets.

One surprising benefit of observing and breeding laboratory rats is an increasing realization by experts that there are similarities in behaviour between all mammals, be they rats or humans. Conscious thought, social communication, childish playfulness, laughter and joy are attributes usually associated with people. Such behaviour is now thought to be common in young mammals as a way to help youngsters weave their way into the social fabric of their kind. Tests have revealed a distinctive series of vocal chirps at 50 megahertz (above the threshold of human hearing) during rough-and-tumble play between juvenile rats, which researchers claim indicates the *'early origins of laughter and joyful social processes that are carried out by the mammalian brain'*.

The common dismissive perception of rats as unclean, detestable vermin underplays their enormous success and significance. Some biologists have even dared to predict a future path of evolution. They have placed the long-term prospects for rats far above those of humanity. If some pandemic or other disaster were to wipe out humans, rats are well placed to adapt. With their varied diets, fast reproduction rates, strong social communities and apparent sense of self-awareness, it has even been proposed that new rodent species could some day evolve into a replacement race of highly intelligent beings (see also crow, page 405).

Australopithecus

FAMILY: HOMINIDAE
SPECIES: AUSTRALOPITHECUS AFARENSIS
RANK: 23

*A fortuitous swivel on to two feet that had
extraordinary consequences*

CHOOSE A SINGLE moment or event that is the most signif-icant to the entire history of humanity – what would it be? The discovery of electricity? The rise of religions? How about the inven-tion of writing? Or, going even further back, the birth of spoken language itself? There is in fact a development more significant than any of these. It probably happened some time between four and five million years ago and occurred within the primate family called the Great Apes (Hominidae). This transformative evolutionary adapta-tion originated in a species ancestral to modern humans, which did simply this: it learned to walk on its own two feet.

The two biggest physiological differences between man and his closest genetic cousins, chimpanzees and bonobos, are that humans have bigger brains (roughly four times larger) and that humans walk on two feet, whereas chimps and bonobos have stayed on all fours.

Australopithecus lived in Africa between about 4.5 and 2.5 million years ago, and since its bones were first discovered by Raymond Dart in 1924 a wide range of skulls, skeletons – even footprints – have been unearthed by palaeontologists. The most famous is a 40 per cent complete skeleton, nicknamed Lucy, painstakingly uncovered in Ethiopia in 1974 by a team of palaeontologists led by American anthropologist Donald Johanson. The shape of the feet, pelvis and spine of this 3.2-million-year-old *Australopithecus* shows beyond all reasonable doubt that her kind walked upright. Volcanic footprints found in Kenya at Laetoli were most likely made by the same, or a similar, species about 500,000 years before.

Why swivel on to two feet? Unfortunately, there is still no scientific agreement on that critical point. There are a number of competing theories. Clearly, there are certain advantages to a life on two feet, rather than four. Stretching up for food that would otherwise be out of reach is one. Looking out for danger in grassland areas where there are no trees is another. Wading through rivers is a third, since it is easier to ford deeper water by walking upright than on all fours. Others think that the gorilla-like ancestors of *Australopithecus* rose on to two feet, Tarzan-like, to ward off predators. It has even been suggested that their two-footed antics could have been a sexual display by males who wished to impress females with the size of their manhood (apehood?), while females rose up to protect their reproductive regions from rearward attack. Some think walking on two feet began as a fashion – something 'cool' designed to impress females (or perhaps pioneered by females to impress males) that then caught on. Once some of the advantages of living on two legs rather than four kicked in, natural selection in favour of those with the best attributes for walking upright took over.

Perhaps the most significant advantage of walking on two feet is the one highlighted by Charles Darwin in his book on human origins, *The Descent of Man*, published in 1871: '*Man could not have attained his present dominant position in the world without the use of his hands, which are so admirably adapted to act in obedience to his will.*' By freeing arms and hands from the task of walking they could then be used for other purposes such as fetching and carrying food for storage to help survive severe weather conditions. When wading across streams hands could grip rocks and trees for safety. Hands not designed for walking could become more dextrous and be deployed for making tools such as clubs and spears to kill fish and other animals for food.

Walking upright is therefore an evolutionary magic bullet – an adaptation that has had a most profound impact on the planet, life and people. Its pioneering species was *Australopithecus*. Until recently it was thought that hand tools originated with the descendants of *Australopithecus*, a line called *Homo habilis*. Traditionally these species are classified as the first true hominids because their brains, at around 650 millilitres, were significantly larger than those of either modern chimpanzees or ancient *Australopithecus*. But in 1996 a research team led by Ethiopian palaeontologist Berhane Asfaw and American Tim White discovered primitive stone tools dating back 2.6 million years scattered next to the remains of a previously unknown *Australopithecus* species. They named it *Australopithecus garhi*, meaning 'surprise' in the local Ethiopian language.

This find suggests that *Australopithecus* gained major survival advantages as a result of walking upright and freeing its hands to make tools. What extraordinary evolutionary consequences lay in wait from this single adaptation! Like two-legged dinosaurs hundreds of millions of years before (see page 126), these bipedal mammals were soon well placed to dominate life on land. But, unlike the dinosaurs, they did it not by brute strength, but by using their hands and with what came next, much larger brains.

By about 2.5 million years ago various species of *Australopithecus* – some slender omnivores, others more solidly built vegetarians – gave way to descendants, *Homo habilis*, whose increasing skill at crafting tools made a vital contribution towards the evolution of more advanced hand–eye co-ordination, better problem-solving abilities and ultimately larger brains. Throwing weapons they had fashioned allowed them to hunt at a distance without endangering their own lives, ensuring that the genetic impetus of their two-legged lineage, pioneered by Lucy and her ilk, powered ever on.

Homo Erectus

> **FAMILY: HOMINIDAE**
> **SPECIES: HOMO ERECTUS**
> **RANK: 13**

The longest-surviving human species with an unprecedented growth in brain power

HOMO ERECTUS **HAS** by far the best track record of any human species to date in terms of survival. He originated in Africa about 2.6 million years ago, yet died out only around 70,000 years ago (see plate 5, top). In contrast, modern humans (*Homo sapiens*, see page 191) have lasted for just 160,000 years so far – less than 7 per cent as long. *Homo erectus* was also the original hominid tourist, the first to explore the world outside its continent of origin. About 1.8 million years ago *Homo erectus* migrated out of Africa to populate parts of Europe and Asia, even reaching as far as Indonesia and China, where numerous remains of its skeletons, skulls and bones have been found (Java Man and Peking Man, for example).

Homo erectus was also the first human species to enhance its environment through the controlled use of fire; evidence comes from the discovery in Africa of charcoal remains dating back 1.5 million years. The remnants of fires have also been unearthed in northern Israel dating back to around 800,000 years ago, and more recently in France and Spain, confirming that this species included the world's first barbecue chefs. Cooked meat releases energy more efficiently and more quickly than raw meat and kills microbes – one of nature's inbuilt checks on population growth – providing better nutrition to satisfy the greater energy requirements of these people's increasingly big brains and boosting populations. Fire also protected *Homo erectus* against wild beasts. This made them king of the predators, eclipsing all other species with their superlative survival skills. Rival humans such as *Homo habilis* who did not use fire are thought to have died out about 500,000 years after *Homo erectus* first appeared.

Without doubt the most notable evolutionary adaptation enjoyed by these people was the dramatic increase in the size of their brains. Early *Homo erectus* skulls show a cranial capacity of about 800 millilitres, roughly twice the size of that of *Australopithecus* (see page 186). Later specimens show how this increased to as much as 1,100 millilitres (approaching the average 1,400-millilitre capacity of modern humans).

One possible cause for both the increase in brain size of this species and its migration out of Africa was climate change. Over the last three million years global temperatures have gyrated wildly between long fits of severe cold and shorter periods of interglacial mildness. Only recently has it been discovered that conditions can sometimes flip from hot to cold and vice versa literally within the space of a human lifetime. Feedback loops in the Earth's temperature regulation systems (such as ice caps, carbon dioxide and methane levels) mean grasslands and forests can be reduced to scrubland sometimes in a matter of decades, not millennia, as was once thought.

Some experts now believe similar sudden climate changes provoked *Homo erectus* to migrate out of arid Africa in search of wetter, more fertile hunting grounds. Essential survival strategies such as crafting better tools for hunting, igniting and controlling fires, eating cooked meat and even working out how to build rafts for coping with rising water levels might never have arisen through generations of slow adaptation driven by the selection of genetic instinct alone. Instead, they all required the use of substantial, impressive, fast-acting rational brain power.

Problem-solving skills, communication between individuals in hunter-gatherer bands and the construction of weaponry were all vital for the survival of this species in the dynamic, rapidly changing environmental conditions of an ice age. The potential for intellectual reason, which had been gradually evolving in mammals since the demise of the dinosaurs, came into its full bloom with the rise of two-legged *Homo erectus*. Working with his hands led to the further enhancement of his cognitive skills. Over time, *Homo erectus* individuals with the best mental skills for working out how to survive in changing environments passed on these attributes to their future generations. And so, over not so much time, average cranial capacity grew.

An additional explanation for the dramatic increase in human brain size was proposed by Charles Darwin in his second most famous work *The Descent of Man*, published in 1871. Sexual selection is the second evolutionary force (after natural selection) that Darwin believed may have helped humans enhance their reasoning powers. With freely available hands, males could prove to females their worthiness as mates by 'showing off' (see also *Archaeopteryx*, page 121; and *Homo sapiens*, page 191). Females may have been most impressed by men that could excel in design and technology skills, such as toolmaking or boatbuilding, and survival techniques, such as lighting fires. These attributes were then more likely to be passed on to the offspring of such well-skilled mates.

Likewise, males may have been impressed by females with desirable domestic skills such as being great cooks or entertainers who danced and sang well. Any daughters they bore with these qualities would be more likely to attract future males. Good toolmakers that bred with excellent cooks, and fancy dancers that mated with brilliant boatbuilders would proliferate. Poorly performing individuals would be generally regarded as unattractive to either sex – who wants their children? According to Darwin, this probably led to a pattern of breeding that favoured enhanced mental reasoning, and performance and handicraft skills. Of course, large males that excelled in hunting (or warding off rival suitors) were highly desired by females too, and their genes were also likely to be passed down the line – which accounts for why *Homo erectus* males were typically 30 per cent larger than females (see also sperm whale, page 173).

No one knows to what extent *Homo erectus* were really cultural beings and if they could communicate or talk in the sense that modern humans do. However, the extraordinary increase in brain size during their two-million-year existence on Earth was certainly caused by a powerful interplay of selection forces, be they the brilliance of brain power needed to survive rapid climate change or the courtship contests of sexual selection.

Homo Sapiens

> **FAMILY: HOMINIDAE**
> **SPECIES: HOMO SAPIENS**
> **RANK: 6**

The beginning of a new evolutionary era
dominated by a single species: man

MODERN HUMANS FIRST emerged in Africa about 160,000 years ago. They evolved from the stock of *Homo erectus*, the long-est-surviving species of humans, whose brains had grown out of all natural proportion (see page 187). By about 25,000 years ago *Homo sapiens* was the only surviving species of human on Earth. According to recent genetic analysis, a near-catastrophic event about midway through its history reduced the population of *Homo sapiens* to no more than a few thousand individuals. Speculations as to the cause of this near-extinction range from viral infection to the giant volcanic eruption of Mount Toba 70,000 years ago that poisoned Africa's habitats. Large brains with a cranial capacity of *c.* 1,400 millilitres may have proved vital to its survival. During catastrophic events advanced mental reasoning powers truly come into their own. Quick thinking, manual dexterity and increasingly complex tool-making skills are likely to have made the critical difference between life and death.

At about the same time the behaviour of *Homo sapiens* started to change significantly. The species began to develop a unique capacity for complex spoken language, learning, education and other forms of what is now called intelligence. No one knows for sure when human language first appeared because, unlike writing, speech leaves no physical record. However, highly sophisticated toolmaking techniques date back around 70,000 years. Spoken language communication would probably have arisen by then since it is difficult to see how such people could have passed on their craft skills from one generation to the next without some kind of verbal communication. Extreme survival conditions may also have put greater emphasis on speech as an aid to hunting as climates changed rapidly during the Earth's oscillating bouts of glaciation.

One possible explanation for the rise of language comes from the same sexual selection pressures that are thought to have increased the rational thinking powers and brain size of *Homo erectus*. With populations of *Homo sapiens* falling to just a few thousand, competition between men to mate with the few remaining women and produce offspring is likely to have become extremely intense. It would have been natural for females to choose carefully from the available men. Those with the most impressive skills would tend to be the most attractive or desirable fathers for their children. With such a small population bottleneck, courtship contests (or, as Darwin called it, 'sexual selection') could have provided the trigger for new advances in which verbal language skills became a male's most potent device for persuading females to mate.

Verbal courtship displays have frequently manifested themselves in spoken and written culture throughout recorded history. Plays, poems, songs and stories about love and romance appear to have supreme importance in human culture, as popular today as they were in the medieval/troubadour age of chivalry hundreds of years ago. Mostly, they are written and performed by men in a bid to impress women, who are frequently influenced by such displays as they determine their partnership choice.

With such small human populations 70,000 years ago, those males that were really impressive linguistically, the most charming perhaps, were most successful with verbally receptive women. Their linguistic skills therefore passed into future generations and the human capacity for language has become mammalian reason's most impressive evolutionary product yet. Complex language, as opposed to simple birdsong, may therefore be best explained as a highly efficient way for male and female humans to find out about each other during the critical phase of choosing a suitable mate.

The same process of sexual selection that may have spawned bigger brains in his ancestor *Homo erectus* could also have helped *Homo sapiens* take what has been called a cultural 'great leap forward'. Sophisticated linguistic, communicative and technical skills helped populations of *Homo sapiens* to recover – whatever the cause of their sudden fall in numbers – and, as environmental conditions improved, these people soon began to spread out of Africa, like their *Homo erectus* ancestors. By 50,000 years ago they had reached the Middle East and fanned out across Europe and Asia, displacing other human species, such as *Homo neanderthalensis*, in the process. Some built rafts and canoes, which they used to paddle across the seas, populating continents such as Australia and the islands of Polynesia, where no previous species of humans had ever lived before. Others, about 13,000 years ago, found a way to cross from the eastern edge of Asia to the Americas, via modern-day Alaska, and within 1,000 years had roamed down to the southern tip of South America (see dog, page 336).

By about 12,000 years ago *Homo sapiens* had reached all continents except frozen Antarctica. The arrival of these intelligent well-armed bipeds with their exceptional capacity for reasoning and communication had a dramatic impact on many other living species, especially those that lived in places where humans had never previously ventured. First in Australia, then in the Americas, mass extinctions of large animals took place soon after human hunters arrived. Large mammals in Africa and Asia had lived for at least two million years in close proximity to humans, enough time to have developed instinctive fear, hiding or fleeing to avoid close contact.

But mammals in Australia and the Americas had no such advantage. Nor had they sufficient reasoning powers to compete with or outwit the brains of man.

As human populations migrated across the globe, evidence of their advanced culture emerged in the form of cave paintings that venerated animals, carefully crafted figurines dedicated to a fertility goddess and ceremonial burials that indicated that this was an increasingly 'self-aware' species. The reasoning powers of *Homo sapiens* were now sufficiently robust to begin to seek answers to questions about the origins and purpose of life on Earth (such as Australian aboriginal Dreamtime stories), life after death (burial ceremonies) and even morals for distinguishing between right and wrong behaviour (such as totems and taboos).

Other physiological traits further distanced this species of human being from its immediate ancestors. So large were the brains of infantile *Homo sapiens* that a child had to be born much earlier than the offspring of other mammals, so that its head could squeeze through the mother's birth canal. These vulnerable human offspring required the constant attention, protection and care of adults. Social groups also proved effective for the transfer of cultural values (especially language and crafts) and other essential life skills, which were taught by older generations to the young. Such birth dynamics may also explain why there is less sexual dimorphism in *Homo sapiens* than among its ancestors *Australopithecus* (see page 184) and *Homo erectus* (see page 187). Males are just 15 per cent larger than females, on average, indicating a greater inclination towards monogamous relationships and nuclear families in which both parents invest time in raising their offspring.

By far the biggest impact of these mostly hairless two-legged hunters with language, art and a curiosity about their place in the world began to be felt about 10,000 years ago. As a result of dramatic swings in global temperatures, traditional hunting grounds began to disappear under rising seas and wild game became scarce. As a consequence humans had to use their brain power to find new ways to feed themselves.

Natufian people, who lived in the area of the Middle East now called Lebanon, retreated into the hills and began sowing the largest seeds of wheat they could find in well-prepared soil. This was the first known attempt by humans to grow crops for harvesting months later. They discovered that grain could be stored in makeshift huts and then used as food, as and when necessary, after being ground into flour and baked into bread. Humans in several parts of the Middle East, North Africa and Asia also figured out, through trial and error, that some wild animals could be tamed and bred into forms that were more suited to the needs of human beings. After being bred in captivity over several generations, such creatures became compliant, their genetic instincts of fear and aggression – vital for survival in the wild – became suppressed, turning them into the domesticated breeds of goats, sheep, cows and pigs that we recognize today. Such animals could be kept on farms – or in herds to support pastoralists – providing regular supplies of milk, cheese, hides, furs and meat. Now *Homo sapiens* had its own 'artificial' solution to hunger management that was more resistant to the effects of climate change and the scarcity of hunting grounds.

The rise of farming had a huge impact on the planet, life and people. So much food could now be kept in store that people could settle in permanent dwellings and many of them were able to explore ways of life beyond the tasks of food provision. The rise of artisans, priests, rulers and soldiers led to the development of large human settlements, well-guarded cities, empires and complex civilizations. Others stuck to the more traditional nomadic lifestyle, taking their flocks and herds with them for food, trading between settlements, or carrying out raids when the opportunity arose. They began to use domesticated animals, such as horses and camels, as personal transport systems, or to carry goods or wage war (see pages 246 and 252).

Humans learned how to convert their spoken language into a

permanent written form using marks and scratches. Such codification further helped pass on knowledge, such as the best time to sow crops as well as keep accurate accounts of trade and exchange. Thanks to cultivated food, animal husbandry and the application of rational thinking, human populations grew from approximately *five million* individuals *c.* 12,000 years ago to approaching *seven billion* by AD 2010.

Human societies differ in one important respect from their counterparts in the insect world (see page 157): they are products of central planning and authority, the work of thinking minds, not of instinctive self-organization that arises through genetic inheritance. Such a difference introduced a radical new selection pressure into the process of evolutionary change. In this world of 'artificial selection' species thrive or die not as a result of their fitness for any given ecological niche but rather because of their usefulness to humans. Over the last 12,000 years global glory has followed those species that have successfully adapted themselves to this new evolutionary paradigm. Isolation has come to species that have failed to prove their worth, many of which have been driven to extinction. A few groups have soldiered on regardless, remaining representatives of those species strong and fit enough to resist the clamour of humanity's rational demands (see On Rivalry, page 395). These, for now, continue to evolve according to the laws of nature – not humanity.

After Humans

**From 12,000 years ago
to the present day . . .**

**On the impact of species that
thrived in the presence of modern
mankind**

9

On Agriculture

How the birth of farming helped create a new top tier of world-conquering species, thanks to the co-evolution of humans and a few select animals and plants.

Wheat

> **FAMILY: POACEAE**
> **CULTIVAR: NORIN 10**
> **RANK: 19**

A highly nutritious genetically modified grass
that stops millions starving

A UNITED STATES law passed on 14 December 2006 declares unequivocally that Dr Norman Borlaug (1914–2009), an American agronomist, '*saved more lives than any other person who has ever lived*'. More than one billion people alive today owe their existence to this man, it is claimed, yet only a tiny fraction will ever have heard of him. What miracles, you may rightly wonder, did this man perform to deserve such unprecedented praise?

In a word, it was wheat. Not just any old variety, but the pioneering of a new hybrid called Norin 10, which since the early 1940s has increased the yield of crops grown in places such as Mexico and India by more than ten times their historic average. Worldwide fears of mass starvation caused by the relentless rise of human populations have been staved off by the cultivation of this carefully crossed grass species that grows on short stalks but has very large seeds. The dwarf-like characteristics of Norin 10 mean it does not topple over despite the weight of its bloated grains. Fed with plentiful supplies of water, artificial fertilizer and chemical pesticides, Borlaug's ingenious wheat pulled back many regions of the world from the brink of disaster in the mid-twentieth century.

Wheat is a species of grass that has truly conquered the world. Its rise to predominance began about 12,000 years ago, when humans in the eastern part of modern-day Turkey began to cultivate varieties of grasses, tending to select those with the largest, most nutritious seeds from their stores. The discovery of how to grind grains of wheat into flour and then bake bread is one of history's most important but least recorded stories (see yeast, page 302).

For early farmers the transformation of wheat flour into a sumptuous leavened loaf was a miracle that lay at the heart of their religious worship. Excavations at Achilleion, a Neolithic settlement near Farsala in eastern Greece, have revealed several ancient temples (*c.* 6000 BC) dedicated to a pregnant goddess. Hundreds of figurines have been found next to bread ovens equipped with platforms on which these statues stood, turning them into shrines and altars. This was the deity that represented Mother Earth with all her glorious bounty, worshipped wherever grain was ground or bread prepared.

Thousands of years later, the association of bread with divinity was not lost on the scribes of the Jewish Torah (Christian Old Testament). They revealed how God sent manna from heaven, a type of bread that was ground up and baked into cakes, to sustain the Israelites on their epic journey out of Egypt to the Promised Land. Two thousand years ago a Jewish carpenter's son called Jesus became, in the eyes of many, the personification of what it took to sustain humanity. He was, he said, the bread of life. To this day millions of Christians celebrate the life of Jesus by eating bread that symbolizes his holy body.

Bread-making has not only permeated religious cultures. Its impact was just as significant in the world of science and engineering. Cogs, sails, mills and wheels were devised by humans from ancient Greece (33 BC) to China (*c.* AD 200), originally for the purpose of pounding and grinding grain into flour. It was only later that the concept of harnessing wind and water power was applied to sawing wood, pumping water, making paper or driving looms. Today's ubiquitous renewable source of electricity, the wind turbine, is adapted from a technology that has its origins in the processing of domesticated wheat.

The amount of land now allocated to wheat cultivation is a staggering 600 million acres – terrain that was once mostly forest or savannah. According to at least one eminent historian, wheat has '*diversified more dramatically, invaded more new habitats, multiplied faster and evolved more rapidly without extinction than any other known organism*'.

Two major factors help explain the enormous impact of wheat. Gluten is a protein that occurs in far greater quantities in wheat than in any other grain or cereal crop. This is what makes it possible to knead ground wheat seeds (flour) with water into an elastic, sticky ball of dough. With the addition of a little yeast (see page 302), the dough ferments and the carbon dioxide gas given off gets trapped by the elastic gluten, expanding the mixture into a capacious, neatly formed, aromatic loaf. Wheat's second remarkable feature is the ease with which it naturally combines with other grasses to make new species. Modern wheat is actually a combination of at least three different grass species whose genes have quite naturally spliced themselves together to make wheat as we know it today. These are einkorn, emmer and *Aegilops squarrosa*. Such genetic agility helped establish mutant varieties with large firm seeds that were subsequently selected by the first human farmers as promising sources of food.

If variety was the key to wheat's rise to global predominance then the modern uniformity of wheat today makes its heyday look increasingly vulnerable. Unlike ancestral varieties, today's world-conquering crops are entirely dependent on humans for their propagation and survival. The seeds of Norin 10 are too large and too firmly attached to the stalk for wind power to act as a dispersal mechanism. What's more, wild wheat seeds gradually sow themselves mechanically by using their 'awns' (bristles) like oars that become erect and then draw together, drilling the grain into the earth. Modern varieties rely completely on humans, usually using seed drills, to manage the process for them. Therefore, ingenious evolutionary techniques for burying fallen seeds two centimetres into the soil have now been lost.

On Agriculture

Cloned wheat cultivars, such as Norin 10, which are typically grown as a single variety, are highly susceptible to infection by parasites, fungi or viruses since they lack nature's most effective defence against disease – biodiversity. As a result, humans have to spray wheat crops with chemical pesticides to combat the threat of harvest failure, mostly regardless of its effects on other parts of the terrestrial ecosystem. Theoretically a virus or parasite could evolve a way of infecting wheat in a manner that humans could not control (see *The Death of Grass*, page 381). Its reward would be a truly global feast. Without wheat billions of humans would inevitably starve. If catastrophe were to strike the human race, then the survival of Norin 10, now one of the most populous types of plant on the planet, would likewise wane through its own inability to reproduce.

Sugarcane

FAMILY: POACEAE
SPECIES: SACCHARUM OFFICINARUM
RANK: 29

*An addictive sweetener that has become the
world's most widely cultivated plant*

PLANET EARTH IS overrun by a species of vengeful plants that go to war with the human race. That's basically the plot of one of the twentieth century's most celebrated science fiction dramas, the apocalyptic *The Day of the Triffids*. Yet John Wyndham's concept is not so very far from the truth. Sugarcane is a type of grass that has, over the last few thousand years, been instrumental in enslaving millions of humans and intoxicating billions more with its sweet, irresistible flavour.

Sugarcane, like wheat, has a complex natural history. It is thought to have been the product of the genetic fusion of several species of wild grasses, making its descent especially hard to pin down. The plant grows quickly, up to six metres tall, and in sections, rather like bamboo (see page 153). First contact was made by people living on the island of New Guinea, in Polynesia, about 10,000 years ago. They discovered that they could cut down this plant's stems and, by chewing its ends, enjoy a fix of sugary, syrupy juice – the plant's natural store of energy intended for use as power for its next year of growth. By selecting those plants with the sweetest stems, these people cultivated canes for thousands of years until eventually a fifth of a sugarcane's stem contained pure sucrose. After grinding out the juice and concentrating it until it became supersaturated, a substance at least as magical as bread crystallized out of the thick syrup. We call it sugar.

Polynesians, with their outrigger canoes, were seafarers. They took their sacred sugary treat with them to Indonesia and the Philippines before reaching the northern tip of India by 8000 BC, where sugar eventually became a fundamental part of the religious history of southern Asia. One day, shortly after his enlightenment *c.* 2,500 years ago, the Buddha was resting under the shade of a tree by the roadside. Two passing merchants saw that he was a holy man and gave him a piece of peeled sugar to eat. This was the first food Buddha had received for seven weeks. So grateful was the Buddha that he anointed these men, Trapusa and Bhallika, and they became his first disciples. Sugar spread with Buddhism as a holy food down through India to Sri Lanka. Westerners got their first taste in *c.* 325 BC when an officer in the army of Alexander the Great described eating a milky rice pudding sweetened with sugar. By AD 50 sugar had reached China, which was the first civilization to experiment with turning sugar into 'rock candy', before being taken overseas by Buddhist missionaries to imperial Japan.

Sugar's biggest break came in the eighth century with the rise of Islam. Of the many products that swirled from east to west with the whirl of Mohammed's people, sugar and gunpowder were surely the most significant. The Muslims learned how to cultivate the crop by cutting back the canes to a stump, then letting them regrow. This way a single plant could be harvested up to ten times before its productivity waned. Mills originally designed for grinding grain were adapted to extract the juice from sugarcane.

The crop's global conquest then progressed via Christian Europe during the twelfth century. Cyprus, Crete, Italy and Spain were all pressed into sugar production by Crusading states eager to avoid trading with their infidel Muslim rivals. Christian cultivation reached its climax on the island of Madeira thanks to Henry the Navigator, a Portuguese crown prince, sugar's most notorious patron. He was the first to combine the mass production of sugar with the enslavement of black Africans, forced to toil on plantations owned by European whites. Initially his plan was to use Muslims captured from Africa to work in Madeira's sugar fields. However, they persuaded him to grant them their freedom in return for sourcing native Africans, who were, they said, the children of Noah's son Ham, whom the Bible had condemned to be 'servants of servants'.

When Henry agreed terms in 1444, it was the start of a most inauspicious pact. Mercenary Muslims and commercial Christians worked together to bind Sub-Saharan Africans into slavery to grow sugar for export. Later that year the first 235 'Negroes' were shipped out of Lagos to work on the cultivation of sugar in Madeira. Over the next 400 years more than twenty-five million Africans were transported west in slave ships to work in the European colonies of the Caribbean, Brazil and North America, mostly to feed Europe's growing fondness for sugar.

The recent human addiction to sugar is a modern menace that is having an impact not so very different from the triffids of Wyndham's tale. His killer plants caused people to go blind. But sugarcane and sugar beet have a more insidious method. They work in cahoots with addictive substances in drinks such as tea (see page 287) or snacks like chocolate (see page 278). The combined effects of sweetness and the kick of caffeine can easily become too addictive for even the most disciplined human appetites to resist (see On Drugs, page 277). Sugar initially reached wealthy Europeans from the seventeenth century onwards through their fancy for sweetened cakes, puddings and pastries, while it was the habit of adding sugar to tea that ensnared the working classes from the nineteenth century.

Sugar, in its most common form sucrose, also has a profound effect on the human body. The full consequences of the high doses now imbibed by humans are still the subject of much research, dispute and debate. A simple but significant impact has been the birth of tooth erosion. Before humans began eating refined sugar there is little if any record of dental decay as we know it today. Yet after sugar became widely available in the nineteenth century, poor-quality teeth became a huge problem, to the extent that some men were refused entry to the Allied armies in World War I – they didn't have enough teeth even to chew their rations.

The sudden introduction of high quantities of sugar into modern diets has had other, more pernicious, effects on the human body. Recent research indicates that high levels of sugar intake in children may be related to a decreased sensitivity to dopamine, a drug released by the brain connected to an animal's ability to learn. What's more, modern epidemics ranging from diabetes to obesity are on the list of ailments increasingly being connected to our modern fixation with high doses of sugar. So addicted has the world become to this ulti-mate species of success that sugarcane is now the most widely grown crop in the world – even more so than wheat – with worldwide production estimated in 2007 at *1.5 billion tonnes*.

Sugar certainly has parallels with the triffids in Wyndham's tale: its sweet success over the last few hundred years has also resulted in the subjugation of many billions of people, be it through slavery, addiction or ill health.

Potato

> **FAMILY: SOLANACEAE**
> **SPECIES: SOLANUM TUBEROSUM**
> **RANK: 43**

A naturally toxic tuber that has successfully adapted into
one of mankind's most nutritious and popular modern foods

VINCENT VAN GOGH painted a scene of peasants eating potatoes by the light of an oil lamp because he wanted to depict the life of ordinary, poor people who had 'earned their food by honest means'. His picture, painted in 1885, reflects the social stigma attached to potatoes that delayed their rise to global dominance for more than 200 years.

Few people today think of potatoes as being fit only for animals or the very poor, since they are now one of the world's most popular vegetables. Even India and China, traditionally rice-growing nations, have finally embraced the potato, the vegetable of choice in the West, because of its high nutritional value and its capacity to grow just about anywhere. In 2007, for the first time, China became the world's largest potato producer.

Potatoes are the fifth most widely grown food crop today, after sugar, wheat, rice and maize. And yet they are naturally poisonous. Quite why or how South American natives began selectively to cultivate these members of the deadly nightshade family over 5,000 years ago is a mystery of agriculture that will probably remain forever hidden. These are plants whose leaf, stem and nutritious reproductive food store (in the form of underground tubers) were designed by natural selection to be highly resistant to even the most persistent of herbivorous attacks. But people living high up in the Andes in what is now Peru desperately needed a crop that could sustain them and would grow at altitudes of up to 4,000 metres. After years of labour (probably a female endeavour) they somehow focused the potato gene pool into growing large quantities of underground tubers that were not only safe to eat but so nutritionally rich that, when eaten in sufficient quantities, they could provide all the nutrients needed for human wellbeing and survival. The potato is still the only vegetable in the world that furnishes, in one staple, all of life's essentials. It's a list that includes carbohydrate, fat, protein, vitamins (B and C) and minerals such as calcium, iron, magnesium, potassium, sodium and zinc.

The sheer range and number of potato varieties found in South America is itself evidence of the gargantuan effort made by these early Andean farmers to find a food that would work as a high-altitude one-stop shop for survival. Thousands of wild potato varieties persist in the mountains of Peru today, perhaps the relics of ancient, long-lost selective experiments conducted by humans in their efforts to transform a well-armed plant laced with poison into a compliant vegetable that tastes great whether boiled, mashed, roasted or fried.

By AD 1500 when the Spanish conquistadors arrived, they found that at least half the natives ate nothing but potatoes. But just when transportation in the form of European adventurers offered this vegetable its biggest chance for global spread, a surprising hurdle which had nothing to do with taste, digestive intolerance or nutritional value, was placed in its path. The invaders weren't interested in these nutritious, easy-to-grow tubers because (a) they were not solid and shiny like gold and silver; (b) food eaten by native savages could never be, in their eyes, fit for civilized European consumption; and (c) potatoes suffered from a crisis of bad branding: they grew underground, were predominantly eaten by animals and were suspected of having something to do with the occult since they were so closely related to belladonna, the deadly nightshade plant, notorious as an ingredient used in a medieval witch's brew. It is the first of several ironies associated with the history of the potato that Spanish conquerors in the 1530s had absolutely no idea that the real treasure they sought lay not with precious metals and gems but with *chicha* – the curious potato vegetable diet of the enslaved, uncouth, savage natives.

More than 200 years elapsed before the potato became widely accepted in any major European country. Until then it was grown mostly as a botanical curiosity. Various rumours abounded about the potato's curse – everything from it being a powerful aphrodisiac (therefore experimented with by Henry VIII of England, desperate for an heir) to a common source of leprosy. Most people agreed that root vegetables corrupted the blood, spread disease and were generally unfit for human consumption. Anyone who ate potatoes, therefore, was making a direct class statement about themselves.

Despite such prejudice, this vegetable struck back. Ireland was the country that became its adopted European home. The Catholic peasantry, ostracised by British Protestant society, found in the potato a source of food that was relatively easy to grow, tolerant of the Irish weather and which provided plentiful fodder for humans and animals alike. In the two centuries leading up to 1840, the Irish population soared from one to eight million, a rise that reflects the successful establishment of the South American tuber as the country's new staple crop.

But, starting in 1847, a devastating famine and blight destroyed Ireland's potato crop, killing more than one million people through starvation, and triggering the mass emigration of millions more to North America and Australia (see also pages 40–41). It is ironic that that the global transplantation of one of nature's most genetically diverse plants was undertaken with so few varieties that a single protoctist pest, *Phytophthora infestans*, could within a couple of years wipe out an entire human generation's crop. The Irish diaspora took the potato-growing and -eating habits of the Irish to the US, which has since become the West's biggest potato-consuming culture. The rise of this fast-food nation in the twentieth century, with its hunger for burgers and fries, is mainly explained by the nutritional value packed into potatoes, which can grow just about anywhere from the plains of Idaho to the mountains of Peru.

By the twentieth century the potato had smashed its way through the last vestiges of European prejudice to take its place at the world's top table. It was a symbiotic relationship – the potato depended on humans for global cultivation and people relied on potatoes to nourish their ever-increasing numbers. After the United Nations declared 2008 as 'the year of the potato', the worldwide conquest of this versatile vegetable was officially complete. The UN now trumpets the potato as a cheap, plentiful and highly nutritious foodstuff that is essential as human populations climb ever higher.

Olive

FAMILY: OLEACEAE
SPECIES: OLEA EUROPAEA
RANK: 80

*A multi-purpose self-preserving fruit that
oiled the cogs of modernity*

THERE IS A saying, attributed to the most holy book of Jewish
religious law, the Talmud, that it is easier to raise a legion of olive
trees in Galilee than to bring up one child in the land of Israel. What
this proverb reveals about olives is extremely pertinent to the story
of this Mediterranean fruit that helped shape the modern Western
world.

Olives grow naturally in Asia Minor. They were among the
first fruits to be cultivated by Neolithic people following the last
ice age because they tolerate craggy soils and don't require manual
cultivation. Other crops, such as barley, were also an important
source of food for those living on the north-eastern shores of the
Mediterranean. But by *c.* 2000 BC many areas were already showing
signs of soil degradation as a result of intense irrigation and farming.
It was the souring of the earth that led to the fall of the Sumerian
civilization in the Middle East.

Olives, as one of the only easy-to-grow sources of nutrition, therefore began to gain the status of a divine gift. From Minoan Crete
to Mycenaean Greece and Israelite Canaan to their Philistine and
Phoenician neighbours, ancient cultures around the Mediterranean
hallowed these fruits as a sacred link between the human and spiritual worlds. Saul, King of Israel, was anointed by God's prophet

On Agriculture

Samuel with a horn of olive oil. The Garden of Gethsemane, where Jesus took refuge to pray on the night of his betrayal, was an olive grove. Even the name Christ means, literally, 'anointed' one, and the Mount of Olives, in Israel, is still the world's holiest burial site for Jewish people today.

A newly planted olive grove takes from eight to twenty years to mature and produce fruit. Once established, however, these trees produce crops year after year, sometimes lasting for more than a millennium. Olive groves need little labour other than cutting back in the winter and vigorous shaking when the fruit is ripe, to gather the olives into nets for processing. Craggy, acidic soil does not bother these trees, which was ideal for people wanting to settle away from fertile, alluvial river valleys.

Such a lifestyle, however, chiefly depended on peace. Attacks that destroyed mature groves were devastating to a way of life founded on trading olives because it took at least a generation to re-establish the crop. Preserving olive groves therefore became a big incentive to resolve disputes via discussion. Trade, compromise and negotiation were the hallmarks of a new style of human society stimulated by olive growing and still symbolized today by the olive branch as a sign of peace.

Olives weren't simply nutritious to eat. Their oil, extracted in giant stone presses, was as valuable to the ancient world's economy as crude oil is today. Not only could olive oil be used in lamps and as fuel for heating, but it was also ideal for cooking, making soaps and cleansing the body. The olive's other natural bounty comes from a symbiosis with a strain of bacterium called *Lactobacillus plantarum* (see page 421). As harvested olives are left to ripen in the sun, this bacterium's fermentation process produces lactic acid, a natural preservative, which allows the fruit to be stored for up to a year without going off. Nothing could be better as tradable produce for people who relied on importing other products (chiefly wheat from Egypt) for their welfare. Civilizations such as the Minoans,

Phoenicians and later on the Greeks all grew wealthy on the trade of self-preserving olives.

Not having to work in the fields all year round gave people in olive-growing cultures the time and incentive to build fleets of ships to further their bartering of olives for grain. The seafaring Phoenicians spread olive growing all around the Mediterranean, to North Africa, France and Spain. Meanwhile, by *c.* AD 650, the Lydians from Asia Minor pioneered a non-violent mechanism for commerce using minted silver coins. This system of exchange, based on tokens, was more easily operated by people with a surplus tradable product. Such commodities provided an additional form of economic security.

Trade in olives also led to the need for skilled craftspeople such as potters. Amphorae were vessels produced to store and transport olive oil. Artisan, merchant, trader, banker and shipbuilder were therefore all professions that arose in Greece to support the olive trade. The contrast with China, where rice growing shaped society into a more rigid tripartite structure – farmers, soldiers and bureaucrats – could hardly have been more marked (see page 236).

The other precious by-product of an economy based on growing and trading olives was time. Since people were no longer shackled to farming all year round, they could engage in politics (democracy, even) or pontificate about the laws of nature and the mechanics of the universe. Thales, Aristarchus, Pythagoras, Plato and Aristotle were just a few of the brilliant thinkers who were fortunate enough to live in a society where food production could be managed by the few on behalf of the many. An olive-based economy therefore underpinned the freedom to investigate and articulate new natural, scientific and rational perspectives on man's relationship with nature.

With the rise of the Roman empire, olives were no longer quite as dominant as they once had been between the time of Minoan Crete (*c.* 2200–1600 BC) and the demise of Greek independence

On Agriculture

(*c.* 160 BC). By 30 BC the bread basket of Egypt was firmly in Roman hands and Europe's dependence on crops grown on one side of the Mediterranean Sea to be exchanged with grain grown on the other was replaced by an enforced one-way flow of treasure in the form of taxes and slaves to Italy.

But the olive's religious and cultural associations lived on – particularly as a sign of peace and divinity. As European maritime explorers spread across the world from the sixteenth century onwards, they took their biological and cultural survival kits with them. The Jesuits brought olive seedlings to California, to ensure they were equipped with the necessary ingredients for holy sacraments such as baptism, confirmation, ordination and marriage (see also grapes, page 307). From there olives spread to South America. Other European expeditions took olives to South Africa, Australia and New Zealand.

Such is the hallowed status of olives today, cultivated since ancient times, that without them it is conceivable that the philosophy, culture and lifestyle of the modern Western world might never have taken seed.

Cod

> **FAMILY: GADIDAE**
> **SPECIES: GADUS MORHUA**
> **RANK: 57**

A robust saltwater fish that fuelled European expansion

IT IS EXTRAORDINARY to think that a creature once as common as cod is now threatened with extinction. For more than 1,000 years, humans living in the northern hemisphere have relied on this species of highly desirable fish to sustain their economic and nutritional welfare. Yet today stocks are so low that the Atlantic cod has joined the inglorious ranks of the world's most vulnerable species.

Viking fishermen found copious stocks of the cod species *Gadus morhua* when they started to explore the Atlantic seas from the ninth century onwards. Between AD 985 and 1011, five Viking expeditions discovered Iceland, Greenland and the east coast of America (Newfoundland). By catching and drying these large omnivorous fleshy fish, the intrepid Norse explorers had a nutritious chewy meal that they could depend on and which didn't easily go off.

Later, Spanish fishermen from the Basque region discovered that by bathing cod in salt they could increase its shelf life still further. Unlike most other fish, cod is unusually low in fat (typically 0.3 per cent), making it easy to salt. Its high protein content (80 per cent when dried) meant that, even in the pre-refrigeration age, this food could be successfully transported thousands of miles back to Iberia and sold on to markets in the Mediterranean for a good price. What's more, after being reconstituted in water, this dried salted fish tasted just as good, if not better, than when freshly caught.

On Agriculture

Cod congregate in areas where warm and cold waters meet – attracted by plentiful food supplies stirred up by ocean currents. From *c.* 1100 to 1500, German tradespeople from the Hanseatic League controlled a cartel to monopolize supplies harvested from the North Atlantic seas off the coasts of Norway and Russia. Set against them were the Basque fishermen, who made a beeline for stocks off the coast of Newfoundland where the Gulf Stream meets the Labrador Current. There, on a series of shallow banks stretching down the eastern seaboard, supplies of cod were so abundant that by the late fifteenth century it was said they could be scooped up in buckets dropped over the sides of boats. By the sixteenth century, British sailors headquartered in Bristol had discovered the location of the Basques' treasure trove. French fishermen soon joined them, and the ports of St Malo and La Rochelle sprang to life on the back of the west Atlantic cod fishing trade.

When in the spring of 1621 a band of Puritans from Britain set off to colonize the north-east coast of America, they landed at a place they came to call Plymouth, to remind them of home. According to a chronicle written by their leader William Bradford (1590–1657), one reason they chose that place to settle was '*chiefly for the hope of present profit to be made by the fish that was found in that country*'. Despite initial difficulties settling in, they were not left disappointed. Within twenty years, the Massachusetts Bay Colony was bringing some 300,000 cod a year to world markets.

It was thought until relatively recently that nothing other than a complete collapse in the 'natural order' could possibly dent supplies. Off-cuts could be sold to labourers, vitamin-rich liver oil to health-conscious middle-class European families, and premium fleshy white fish to markets throughout America, Europe and the Mediterranean. As a result, the American city of Boston grew rich off the profits of the cod trade, symbolizing its wealth with a gilded cod that hung from the ceiling of the city's prestigious town hall. But only the most exceptional wild species have been able to thrive despite the combined onslaught of increasing human populations and the industrialized food processing that emerged during the nineteenth century. Cod stocks fared well enough until the arrival of steam fishing ships at the beginning of the twentieth century, despite relentless harvesting by man. When these vessels were upgraded to more powerful diesel trawlers from the 1930s, it became possible to drag fishing nets along the bottom of the ocean floor, scooping up all forms of sea life. Valuable cod were harvested in their millions, while the marine ecosystem was devastated, often beyond repair. Meanwhile, freeze-drying, a technique for processing frozen food pioneered by American inventor Clarence Birdseye (1886–1956), further expanded the market for cod in the form of fish fingers, now made available to areas far away from the coast.

Warning signs that cod stocks were on a slippery path to decline came from a series of three so-called 'cod wars' between Iceland and the UK from 1958 to 1976. Desperate to preserve the wealth of its only viable material export – fish – Iceland extended the six-kilometre fishing zone around its coastline to a 322-kilometre zone in 1976. At the same time, enormous bottom-trawling factory ships equipped with onboard sonar systems for detecting shoals of fish with pin-point precision were added to fishing fleets all over the world. Modern factory ships, capable of staying at sea for months at a time, sucked up and processed so much cod (and all manner of other, less economically valuable fish) that by the 1990s global fish stocks faced complete collapse.

Today, there is no guarantee that, despite recent fishing restrictions, *Gadus morhua*, the Atlantic cod, will ever re-establish its previous ecological pre-eminence.

Pig

FAMILY: SUIDAE
SPECIES: SUS SCROFA
RANK: 48

*An unfussy mammal most usually kept
in unnatural conditions*

WHAT PLEASES PEOPLE about a pig? They are not bred for their colour, friendliness or as a source of labour, transport or milk. Nor is there any selective advantage for a pig with brains – even though these are among the most intelligent mammal species and quick to learn. Pigs are popular with most people for one reason only – they taste so good. The Romans summed it up in a saying: '*The race of pigs is expressly given by nature to set forth a banquet . . .*'

Today, pork is the most widely eaten meat in the world. More than a hundred million tonnes of pig meat is farmed, mostly in intensive warehouses, despite taboos in at least two major religions, Judaism and Islam, against its consumption. But by far the most prolific consumers are in Asia. More than half of all pork is reared and consumed in China. Compared with our treatment of other domesticated species, the relationship between people and pigs is a lot more sour than sweet. Pigs were among the first creatures that *Homo sapiens* discovered could be kept in captivity and bred for human consumption. Genetic studies reveal two main areas of initial domestication – in Asia and the Near East – beginning about 9,000 years ago.

Traditionally the life of a pig wasn't so bad. These are natural foraging creatures with an undiscriminating diet. Like humans they are omnivores, and will eat almost anything from dead insects and worms to tree bark, roots, fruits and flowers. Their most powerful tool is their snout, which not only houses a superlative sense of smell but is used to turn over the ground. Humans that tended pigs (swineherds) traditionally allowed their herds to forage in their natural woodland habitat, enabling these social animals to look after their young and move about in family groups. But rearing pigs this way requires a great deal of space because of the potential environmental devastation resulting from their omnivorous diet. Pigs also lack sweat glands so in hot climates they wallow in mud as a way of keeping cool. This only adds to the damage they cause if kept too closely confined, which is why a typical pig sty looks so bare and dirty. With widespread deforestation since Roman times as a result of growing civilizations, demand for pork combined with a lack of space has led to a radical change of approach in the way pigs are kept.

Today pigs are mostly reared indoors on intensive farms. Hundreds of animals are kept in individual stalls where they can be fed on a carefully controlled diet of grain laced with antibiotics, vitamins, hormones and dietary supplements to ensure they mature as fast as possible and fatten up sufficiently to turn a good profit. Females are typically kept in crates, too small for them even to turn around, where they give birth to their litters of five or six sucklings at a time. Their offspring are removed immediately after weaning. Their living space is kept deliberately small as a conflict-avoidance measure since in this most unnatural environment, hostilities between animals often break out, ranging from tail biting to genital mutilation. Castration, tail docking and ear notching are human interventions designed to alleviate some of the symptoms of intensive farming, but they do not address the causes of the obvious suffering of pigs, or the 'misbehaviour' of these creatures. Despite regulations to try to improve pig welfare, there is currently little enforcement of the rules. A recent survey found that more than 80 per cent of pig farms across five European countries were engaging in illegal practices.

On Agriculture

Historic human disdain for pigs is echoed in a legacy of linguistic adaptations. Words such as swine, hog, sucker and runt indicate how little is ever said in favour of the pig. Some cultures (possibly originating in ancient Egypt) have declared pigs so 'impure' (Leviticus, chapter 11, verses 7–8) that they are unfit for humans to touch, let alone to eat. Yet, despite these taboos and such verbal and physical abuse, humans have over thousands of years invested huge amounts of time, money and effort into rearing pigs in ever-increasing numbers. From the point of view of the pig's genes, this represents a considerable triumph. What's more, humans were directly responsible for substantially increasing the range of these creatures far beyond their natural habitats across North Africa, Europe and Asia. Spanish conquistador Hernando de Soto (1497–1542) regarded these animals as being so important for survival that when he ventured to Florida in 1539 with 620 men and 220 horses, he also took with him 200 pigs. During his three-year progress through the southern states many of the pigs escaped, seeding populations of wild boar, now called razorbacks. These creatures still roam the southern United States, where they are hunted by humans for sport.

Today, the supremacy of farmed domesticated pigs as a source of human nutrition is unsurpassed. In China consumption of pork increased by 25 per cent between 2002 and 2007 alone and shows no sign of slowing. As 211,000 humans are, on average, added to the world's population every day, the success of pigs, like that of most domesticated species, is inextricably tied to the fate of humans. Yet, unlike many other domesticated creatures, pigs have proved themselves sufficiently adaptable, intelligent and unfussy about where they live and what they eat to thrive successfully in the wild (see also camels, page 252).

Sheep

> **FAMILY: BOVIDAE**
> **SPECIES: OVIS ARIES**
> **RANK: 26**

One of the most populous domestic species that provides human civilizations with fibre and food

SHEEP WERE AMONG THE first creatures selectively bred to help improve the human way of life for the simple reason that they proved to be nature's most easy-to-manage animals. Their pliability in the face of unorthodox human expectations explains how a thinly dispersed wild animal that probably originated in Asia has now become a ubiquitous creature that numbers more than *one billion* individuals today. Sheep now thrive in vast numbers on every inhabited continent of the world.

The sheep's success lies in its gene pool. Over thousands of years this creature's genes have been focused by pastoralists to produce a suite of products that have changed world history. Wild sheep do not produce much wool. Instead, their bodies are covered with hairs. The fleece of pre-domesticated sheep provided just a thin undercoat of thermal insulation.

In a world full of humans, the most easily controlled animals are those that naturally live in communities with social structures where the majority submits to the authority of a few dominant individuals. If the leadership role is taken over by humans, the species can easily be domesticated. Other characteristics for successful domestication include being non-territorial. The natural compliance of sheep has served them very well indeed. Farmers from the southern Himalayas were among the first to capitalize on these attributes. After many generations of living in close proximity, sheep that prospered in the company of humans were those with the thickest fleeces. Humans preferred these animals because they could pull out clumps of wool with their bare hands and compress the fibres to make felt. They later learned how to shear off the fleeces and spin the fibres into yarn which, using ancient basket- and fence-making techniques, they wove into cloth. Wool rapidly became one of humanity's most widely traded and valuable commodities.

What oil is to the smooth functioning of the modern global economy, wool was to the ancient world. The Sumerians, widely regarded as the first people to build substantial urban settlements, controlled massive flocks of sheep, which provided the wool to clothe most of their people. Wearing warm garments made it possible from about 3000 BC for the Sumerians to live in higher, more mountainous areas, taking their flocks with them, their way of life fanning out across the Mediterranean and into Europe.

Sheep are uniquely easy and cheap animals to farm. They do not require feeding in winter (unlike chickens or cows); they can graze in pastures all year round without needing shelter; and they are able to survive well in cold mountainous areas and in rough, uneven terrain. In return they provide an enormous range of useful commodities: ewe's milk can be drunk or made into cheese. Sheep meat is a major source of nutrition in many parts of the world in the form of either lamb or mutton. Lanolin is a naturally occurring grease extracted from the sheep's fleece that is used in everything from cosmetics (such as Oil of Olay) to industrial lubricants. Finally, wool is a material that has clothed humans from China to the Americas, at least until the rise of artificial fabrics after World War II.

Sheep were central to the economy of medieval Europe. Monastic orders that originated in France and Italy spread to England following the Norman Conquest in 1066. Forests were felled by Cistercian monks to provide pasture for their grazing sheep. Thanks to a papal dispensation that freed these monks from paying taxes, a profitable arrangement was established in which wool grown in the UK was traded in Continental markets to be spun and woven in Flanders and northern Italy. From the eleventh century onwards wool became the most significant international currency for trade with Islam, swapped for gold from the south Sahara Desert in the markets of Champagne, now in northern France.

Families such as the Medici in Florence became fabulously rich, patronizing the arts and triggering what is known to history as the Italian Renaissance. Meanwhile gold from the wool trade filled the coffers of the Cistercians, who became so rich that in the sixteenth century the English monarch Henry VIII used his marital dispute with the Pope as an excuse to avail himself of their fabulous establishments and wealth. The money he gained helped pay for England's first powerful naval fleet – the initial step towards a mighty British empire which, at its height, encompassed 35 per cent of the world's landmass.

Sheep made their biggest modern economic impact on the British-led industrial world. Incentives to find ways of speeding up and automating the manufacture of woollens lay behind some of the Industrial Revolution's most powerful inventions. John Kay's technique for weaving cloth, the flying shuttle (pioneered in 1733), inspired other inventors such as James Hargreaves to find a way of supplying looms with more thread. His spinning jenny (invented in 1764) could spin yarn ten times faster than traditional medieval-style spinning wheels. Later that century ingenious minds adapted water power – and then steam power – to automate the weaving of cloth. Mills were opened wherever natural conditions best suited their location – usually next to fast-flowing rivers – turning the English Pennines into a worldwide centre for cloth manufacture. With demand to feed mills with raw materials outstripping supply, the Highlands of Scotland were cleared of their traditional lairds and replaced by a mostly absent English aristocracy, who dedicated their land to supporting permanent flocks of millions of sheep. The Highland Clearances of the late eighteenth century paved the way for mass emigration of Scots to the Americas, Australia and New Zealand, a fact not forgotten by those still clamouring for Scottish political independence today.

By the end of the nineteenth century Britain was undergoing its irreversible transformation from a mostly pastoral society based on rearing sheep to a series of linked urban metropolises. Wool from India, Australia and New Zealand supplemented the never-ceasing appetite of Britain's steam-driven 'satanic' mills.

Long-lasting, waterproof, insulating woollen fibres are less important today than they once were thanks to the growing ascendancy of artificial fibres in the last sixty years. Nylon, acrylic and polyester have come to replace wool, cotton and linen in many parts of the modern material world.

But sheep have versatile genes. Today they are reared more for their meat than for their wool, and eating lamb is on the rise. Sheep are the only domesticated species that has not succumbed to the economies of scale of intensive farming. These creatures are so simple to farm that families all over the world are still able to scrape a living through the sale of wool and meat, providing their flocks have access to sufficient supplies of rough and ready pasture. Such adaptable creatures have been fundamental to the rise and sustenance of human civilizations, past, present and future.

Cow

> **FAMILY: BOVIDAE**
> **SPECIES: BOS PRIMIGENIUS**
> **RANK: 17**

*Versatile animals which now pose a serious
threat to the global environment*

JACK'S MOTHER WAS beside herself with fury when her young
son announced his latest and greatest deal. On the way to market
the boy had been persuaded by a traveller to sell the family's precious
dairy cow in exchange for a handful of beans. His mum was right to
be cross. The cow was their main source of income and should have
fetched a high price. How could she have known that the beans would
later grow into a giant beanstalk that would help fearless Jack retrieve
the treasure of a miserable giant who lived high up above the clouds?

This nineteenth-century English fairy story reveals the intimate
relationship that has evolved between humans and cows. It's an association that stretches back at least 8,000 years to when wild aurochs,
which originated in India about two million years ago, began to be
domesticated by humans in the Near East. Domesticating a beast
like the aurochs was no mean achievement. In the wild these animals
sought safety from attack by wolves by gathering into vast herds,
fleeing at the first sign of an approaching predator. Contrast this
with the scene of a pastoral milkmaid, approaching confidently with
her three-legged stool. She pats the cow on the back, places her stool
on the ground beneath its giant udders and proceeds to squeeze
them rhythmically until the compliant cow lets down her milk and
the maid's pail is filled.

On Agriculture

No one really knows how such an extraordinary transformation in animal behaviour was actually achieved. Domestic dairy cows are the same species as wild aurochs, but over thousands of generations of living with humans these creatures have proved pliable enough to provide man with a wider range of goods and services than any other living creature. The success of cows is reflected in their present populations. Estimated at 1.3 billion individuals, these are the most well kept, highly bred, populous domestic farm species of all time.

It is thought that the first domestic cattle emerged in Mesopotamia. They were farmed by the same people who pioneered the husbandry of sheep for their wool and milk (see page 224). Perhaps they persuaded the wild aurochs to approach their human settlements by leaving supplies of salt and water nearby as bait. As the animals ate the salt, these Neolithic humans may have approached and stroked them, teaching them to have nothing to fear from the presence of humankind. Charles Darwin revealed that instinct is an inherited trait, and once fear was bred out of their system, compliance in cows took its place.

Cows became so important to the wellbeing of some societies, for example in India, that they are regarded as sacred animals (see plate 8) – provided by the gods as a source of milk and labour – and far too precious to be slaughtered for meat. In Asia, castrated male humped cattle, commonly known as oxen, have been the primary source of automotive power for pulling carts and ploughs for thousands of years. '*Protection of the cow means protection of the whole dumb creation of God,*' declared India's early-twentieth-century independence leader Mahatma Gandhi. *Ahimsa*, the Hindu, Jain and Buddhist doctrine of non-violence to living things, was largely derived from, and most powerfully symbolized by, the sacred status of the domesticated cow.

Different breeds of cow came to have huge significance in other civilizations too. China's ability to feed its massive rice-dependent population owes as much historically to the muscle power of its herds of patient, mild-tempered water buffalo as to its pioneering development of the wheeled metal plough. There are more than 145 million buffalo still collaborating with humans in China today. Tibetan civilization is largely dependent on a single breed of shaggy, thick-coated domesticated cow known as the yak. There are an estimated five million of these pack animals, which provide milk, meat, hides, hair, underwool, dung (for fuel) and blood, which is drunk as a source of protein. Butter from yak milk is even used as fuel for Buddhist lamps. Hair and hide are turned into ropes, saddlebags, blankets, bags, tents, footwear, bottles and boats.

In the Middle East, Europe and Africa the idea of eating beef was never taboo. Cows were the perfect multi-purpose domesticated animal and could be self-transported wherever there was dry land. In Britain wild aurochs were mostly replaced by domesticated breeds before the Roman era. The last European specimen died out in Poland in 1627. Cattle were fully exploited as a source of wealth in other parts of the world during the European colonization of the Americas and Australia. Columbus brought cows with him from the Canary Islands on his second transatlantic voyage in 1493. South American herds were seeded by the arrival of one bull and five cows with Portuguese settlers who landed in Uruguay in 1555. Two hundred years later there were so many herds roaming through Argentina that the government declared any individual could help themselves to up to 12,000 cattle each.

By the eighteenth century, specialist breeding techniques, pioneered by Englishman Robert Bakewell (1725–95), had begun to revolutionize cattle herds. Rather than developing breeds suited to specific locations or lifestyles, Bakewell deliberately inbred animals to increase their yields for producing beef and tallow (for making candles). He selected shorter-legged individuals that matured early and put on plenty of fat, stating that his aim was to produce a breed with '*the greatest value in the smallest compass*'. In so doing, Bakewell began a trend of breeding animals to maximize commercial profit. Success was swift. As a result of his techniques, the average weight of a castrated British bull destined for slaughter more than doubled between 1700 and 1786, from 170 to 380 kilograms.

In the second half of the twentieth century, techniques for producing bovine super-breeds accelerated even more dramatically. Thanks to the practice of artificial insemination, which is used almost exclusively in modern herds, a single pedigree bull can now father thousands of cows without physical contact. Farmers no longer have to keep bulls. Nowadays, mail-order sperm guarantees a herd's thoroughbred credentials.

Some of the environmental consequences of planet Earth playing host to 1.3 billion cattle, such as deforestation, have been obvious for many years. However, it is only in the last forty years that experts have become aware of the giant contribution that cattle make to global warming. A report published by the United Nations in November 2006 revealed that 18 per cent of greenhouse gas emissions come directly from the bellies of domesticated livestock – that's 5 per cent more than total worldwide emissions from global transport systems (road, rail, sea and air). Cows are by far the biggest culprits because of their huge numbers and their penchant for producing nitrous oxide and methane, far more powerful greenhouse gases than the carbon dioxide produced by human industry.

There is a modern moral to be read into the old story of Jack, the boy who swapped the family cow for a handful of beans. Turning vegetarian is probably the biggest single personal contribution any one person can make in today's battle against global warming. Man's 12,000-year-old obsession with the wealth, taste, company and labour derived from domesticated cows is now a big factor in what is fast becoming a more vulnerable, volatile and significantly hotter world.

Chicken

FAMILY: PHASIANIDAE
SPECIES: GALLUS GALLUS
RANK: 61

*How cocks and hens were ideally suited to
the machinations of modern fast food*

BIRDS HAVE NOT featured highly in our list of one hundred species that have changed the world. *Archaeopteryx* (see page 121), widely regarded as the ancestor of feathered flight, had a major evolutionary impact as the creature whose descendants came to dominate the skies. Birds were also remarkable survivors of the catastrophic meteorite collision 65.5 million years ago that smashed the rest of the dinosaurs into extinction (see page 125). Today only one avian class survives (out of an estimated eight that evolved following the emergence of *Archaeopteryx*). They are called the Neornithes. All birds – from ostriches to ducks and eagles to swallows – belong to this single class, but none have had as big an impact on the planet, life and people as *Gallus gallus*, the domesticated chicken.

The chicken's rise to dominance began somewhere in the Far East about 8,000 years ago. They were originally bred in places such as India as much for cock-fighting as for their meat and eggs. Indeed, the most popular meat breeds today have their origins in varieties such as the British Game Cock, bred primarily for the broad-set breast and strong legs that made it an excellent fighter. This is where the boxing terms 'bantamweight' and 'featherweight' originate. As early as the fourth century BC, Alexander the Great's troops brought fighting cocks back to Europe from their campaigns in India, although the first evidence of chickens becoming an important part of the Western diet goes back to the Romans, who fattened their chickens on a diet of milk and bread.

Roman legions then brought chickens to Britain. They soon became the bird of choice in Anglo-Saxon settlements, primarily as an egg-laying farmyard favourite. Meanwhile, Polynesian explorers – descended from Chinese farmers of the Yangtze valley – spread the bird farther eastwards across the Pacific, as far as Easter Island and even to the coast of the Americas. The recent discovery of chicken bones that date back to the pre-Columbian Americas is shedding new light on pan-Pacific expansion. Canoe-paddling Polynesians from Hawaii and Easter Island, it seems, must have made the extraordinary journey to the west coast of the Americas decades before the arrival of Spanish adventurers. DNA analysis of ancient chicken bones in El Arenal, Chile, dates them back to the fourteenth century, before contact with Europeans was established. Vikings to the east and Polynesians to the west were, it seems, the real discoverers of the Americas, long before Columbus set sail.

Chickens were traditionally bred for cock-fighting in the UK at least until the mid-nineteenth century. At one time the sport was as popular as football is today. Cockerels were consigned to a pit where they would literally fight to the death. As in ratting (see page 182), human observers would take bets on which bird would become champion. With the opening up of China to international trade from the 1840s chickens became a more popular source of food and nutrition. European 'Hen Fever' was fuelled by Queen Victoria's fancy for chickens. She even had a poultry shed built to house a flock of Chinese Cochins. The keeping and breeding of exotic chickens for their looks and the curious brown eggs they sometimes laid had the cachet of royal approval. The age of the chicken show had arrived.

But it was not until the twentieth century that chickens swooped up the scale of those species that have most changed the world to become what is by far the world's most numerous bird. Today there are an estimated *twenty-four billion* chickens worldwide – that's more than three birds for every person on the planet. Most are bred for their meat, although egg production is also big business. Chicken is now the cheapest source of meat in the world and has dominated the cuisine of all human cultures from China to the Caribbean.

Mass-production techniques, pioneered by American Eli Whitney (1765–1825) in his water-powered armaments factory, proved ideal in the rapidly growing business of intensive chicken farming. Broiler belts automated the killing, plucking and butchering of birds so they could be sold in faraway markets, ready to cook. The discovery that, with a suitable vitamin D supplement, chickens could be kept indoors twenty-four hours a day, cooped up in purpose-built factory sheds without natural sunlight, was a second breakthrough that came in the late 1920s. The less the birds moved around, the less feed they needed and the faster they grew. Techniques like these, begun in the US, were soon transported to Britain and beyond. With rationing of beef and pork during World War II, home-grown chicken meat and eggs became the West's favourite and cheapest foods.

But it was in the post-war era of the twentieth century, when American mass-production techniques first arrived in the UK, that the chicken really came into its own. The six-week life-cycle of a battery chicken, which never sees the light of day, compares with the twelve-year lifespan of a natural chicken allowed to feed in the open and not killed prematurely for its meat. The extraordinary capacity of domestic chickens to produce meat and lay eggs in the highly cost-effective environment of intensive farming provided essential fodder for the fast-food industries of the developed world. In 1954 'Colonel' H. Sanders (it was a self-styled title, he was never in the military) opened his first Kentucky Fried Chicken outlet in Salt Lake City. After conquering the Wild West, cheap chicken fast-food outlets made their first complete circuit around the USA before arriving in the UK in 1964. So successful was the instant-chicken-eating revolution that by 1983 the McDonald's burger chain had launched its own range of finger food in the form of Chicken McNuggets. Since then, McDonald's has upped its retail presence from 6,000 to 30,000 restaurants worldwide.

Increased meat consumption means increased demand for chicken feed. Soya beans (*Glycine max*) are one of today's most widely grown crops, largely thanks to the need to feed the world's chickens. Livestock feed now accounts for up to *half* of the world's entire agricultural produce, of which chickenfeed represents the lion's share. The impact on the planet concerns the scale of agricultural support needed to maintain this trend. A sustainable response to the ever-rising human population would be for each person to reduce their average meat consumption and for governments to incentivize such behaviour. The reality today is the reverse. The wealthier people become, the greater their lust for meat – life's least environmentally efficient source of essential protein.

Gallus gallus represents the apex of human/animal domestication. This once exotic wild Asian guinea fowl is now the most populous domesticated species on Earth.

Rice

> FAMILY: POACEAE
> SPECIES: ORYZA SATIVA
> RANK: 27

A miracle crop that sustains billions

HIGH UP IN a remote land-locked mountainous region of the Philippines, ten hours by bus from the islands' capital Manila, lies what many regard as the eighth wonder of the world. Ancient tribal people constructed this marvel – a vast series of stepped agricultural terraces that jut out from the steep mountain slopes. The terraces, built about 2,000 years ago, range across an area totalling more than 10,000 square kilometres. The reason these people went to the enormous effort of building fields out of mountains can be summarized in a single word: rice. Island environments are precarious at the best of times. When ancient human settlers arrived in the Philippines from Taiwan they knew their welfare depended entirely on cultivating the crops that they had brought with them. Rice, first domesticated along the Yangtze river valley in southern China, was and still is the most nutritious foodstuff ever known to man. Today more than 2.3 billion Asians – nearly half the world's human population – depend on this single crop to provide as much as 80 per cent of their daily intake of food (see plate 7, top).

Chinese people first began to domesticate rice between 8,000 and 5,000 years ago, at about the same time that Neolithic farmers in Mesopotamia were taming their first sheep (see page 224) and cultivating crops of wheat (see page 202) and barley. Like emmer wheat, wild rice is not genetically suited to agriculture. Its seeds shatter when ripe, scattering them as far as possible in the wind, making it hopeless for humans to gather as food to eat. The grains are also hard and gritty; nature designed them to survive long periods of dormancy on the ground, equipping them with a tough outer coating that protects them until the arrival of damp conditions when germination can begin.

But, like a few other select species, wild rice had a sufficiently diverse pool of genes to adapt in ways that eventually pleased people. Careful selection and cultivation by early Chinese farmers produced varieties with seeds that stuck to the stalk, and soft grains that were permeable to water. These could be cooked in boiling water, quickly releasing a plentiful source of essential edible carbohydrates and protein. So rich was this nutritional prize that Asians soon learned how to grow the crop for best results. They developed elaborate irrigation systems. They found that waterlogged paddy fields kept pests and weeds at bay. They sowed each plant in regimented rows so they could manage and harvest their crops more easily, increasing the yield. Advanced agricultural skills in growing rice largely account for why, until about the eighteenth century, the people of China led the world in their knowledge of agricultural science and innovation. The blast furnace, improved in AD 31 by court scholar and inventor Du-Shi, with its water-powered bellows, led to the manufacture of a new range of agricultural tools including hoes and adjustable ploughshares that could make heavy clay soils suitable for growing rice.

Human populations that live on rice tend to be more numerous than those that feed on wheat or olives because rice cultivation is highly labour-intensive. Like sugarcane, each plant has to be sown individually every year and fields have to be properly ploughed and irrigated at the optimum time to ensure a good harvest. Until new varieties of 'miracle rice' were bred by scientists in the 1960s one hectare of rice could sustain 5.63 people for a year. This compared with just 3.67 people per hectare for the Western world's rival, wheat. The effects on population growth between rice and non-rice cultures were apparent by 1750 when the total population of Asia reached 502 million, more than three times that of Europe. Big populations meant big governments and big armies, quickening the conquest of land by Asian rice. For the last 5,000 years forests have been uprooted for paddy fields and new rice varieties have been cultivated that require less water and can grow at higher altitudes. By 2,000 years ago, some rice-growing people had migrated from mainland Asia to Pacific islands such as Taiwan and the Philippines, taking their precious grains with them.

With the rise of Islam from the eighth century, Asian rice spread throughout Europe and North Africa, along with a clutch of other significant Chinese inventions, such as gunpowder, paper, wheel-barrows, compasses, umbrellas, stirrups, matches, playing cards, kites and chess. Human tastes in the Middle East were re-educated to accommodate this new rich source of nutrition. Spanish paella, Italian risotto and the paddy fields of the French Camargue are all traditions that continue to this day, but began in ancient China and were later transplanted to Europe. It happened thanks to the universal appeal of rice, the relentless zeal of Islamic traders and their enduring conflicts with Christian crusaders, who took rice-growing even farther west.

Meanwhile, in west Africa a related but different species of rice, *Oryza glaberrima*, was domesticated by natives living around the Gambia and Niger rivers. Africans developed their own expertise in rice cultivation, mirroring the irrigation techniques found in ancient China. Portuguese sailors who first explored the west African coast in the fifteenth century were astonished at the agricultural riches they found. Specialist long-handled shovels called *kajandu* were used by the natives to turn over the earth and prepare their paddy fields. European explorers brought with them the Asian rice species, *Oryza sativa*, which has now largely replaced the African type.

Expertise in rice cultivation travelled with African slaves shipped

On Agriculture

across the Atlantic to Charleston to work on American sugar, tobacco and cotton plantations from the seventeenth century. Those with knowledge of how to farm rice fetched the highest prices at auction. Until the abolition of slavery after the American Civil War (1861–5) Asian rice was one of the most profitable and widespread crops grown in the southern United States. It is still grown today along the banks of the Mississippi river. Rice cultivation in California caught on when Chinese labourers flocked there during the years of the gold rush (1848–55) taking their rice-growing skills with them. Rice had completed its circumnavigation of the globe.

Following World War II, when it seemed that many parts of the world were on the brink of starvation, it was to rice (along with wheat, see page 202) that scientists turned for a way of avoiding a great Malthusian famine. Painstaking cross-breeding techniques pioneered by experts at the International Rice Research Institute (IRRI), appropriately located in the Philippines, led to the development of a new variety – IR8 – with dwarf-like characteristics that could tolerate the higher yields generated by artificial fertilizers. When this variety was made available to the world in the early 1960s it was dubbed – for good reason – 'miracle rice'. Yields shot up in some places to as much as 10.3 tonnes per hectare, up from the average one tonne per hectare produced by traditional varieties. What's more, IR8 matured in just 130 days instead of the normal 170 days. Later improvements (breed IR36) have cut the growing time to just 105 days, allowing many farms to cultivate two crops per year – more than doubling their output.

The result is that instead of famine and declining populations, the last fifty years have witnessed a rice binge that has boosted the number of people from 2.5 billion in 1950 to nearly seven billion today. By far the biggest rise is accounted for by Asia – where rice eating predominates – up from 1.4 billion to 3.9 billion over the period. All this leaves today's world with an almighty conundrum. Feeding ever-rising human populations will require a step-change in agriculture – as predicted by Norman Borlaug (see page 202). Since there is little uncultivated land left to press into new production, increases in supply must come from greater yields. Unfortunately, the yield of modern rice varieties is fast approaching biological limits, regardless of how much nitrogen fertilizer is dolloped on to fields.

Maize

> **FAMILY: POACEAE**
> **SPECIES: ZEA MAYS**
> **RANK: 35**

A super-photosynthetic grass with sinister side-effects

HUMAN SACRIFICE, CANNIBALISM, vampires and obesity are horrors best left on the top shelf of Hollywood's library of 'video nasties'. Unfortunately, for people and animals alike, these terrors do not exclusively belong to the realms of fiction. They are also linked by a type of grass that dominated much of the Americas until about 500 years ago, when it suddenly spread across most of the rest of the world. Since then, this crop has become one of the three most widely cultivated, and potentially most lethal, species on Earth. American farmers call it corn, although to the rest of the world *Zea mays* is better known as maize.

Quite how natives living in Central and South America bred wild grass of the *Zea* family into a species with an enormous ear (a cob) that contains a cluster of up to 300 highly nutritious edible seeds (corn) is baffling. Despite extensive research, genetic sequencing and intense scrutiny by countless archaeological experts, there is still no single convincing theory of how fast-growing teosinte, a wild spiky grass that hosts between five and ten inedible seeds, morphed into maize with its remarkably succulent, highly digestible cob. Certainly the process took thousands of years. Evidence of early domestication reaches back at least 5,000 years, around the Balsas river valley in Central Mexico, where wild teosinte still grows. A mixture of human selection and whimsical nature best accounts for how hybrids fused

241

into a new super-species, *Zea mays*, with its highly complex genetic structure. Maize has more than twice as many genes as humans.

From about 1500 BC maize cultivation began to transform the lives of people living in Central and South America. New civilizations emerged that developed their own systems of government, writing, crafts and exchange. Since these regions lacked the domesticated animals of Africa and Eurasia (see sheep, page 221; cows, page 228; dogs, page 336; and pigs, page 221) they became almost totally dependent on crops such as maize. In times of drought, sacrifices were most often made to Tlaloc, the Aztec god of water and rain. Women and children were primary victims, their tears symbolizing the plentiful rains that Tlaloc would surely bring to water these desperate people's fields of maize.

Dependency on maize may also explain why cannibalism appeared on the menu of maize-dependent Native Americans living in the southern states of North America. Maize does not provide the full range of proteins, vitamins and sugars required for a healthy human lifestyle (unlike potatoes, see pages 210); meat is an essential addition. Anasazi people spread the practice of maize cultivation throughout the southern states of North America from *c.* AD 600 until *c.* 1150. Over-zealous deforestation (in order to grow maize crops) and a prolonged drought starting about AD 1130 led to their demise as they abandoned their holy city of Chaco Canyon soon after 1170. In a land devoid of domesticated animals and left with precious little game, the decision to eat human meat may have been a nutritional necessity. But it was not sustainable in the long term since, generally speaking, humans are reluctant to allow themselves to be domesticated as farm animals for consumption.

The arrival of European colonists in the early sixteenth century following the voyages of Christopher Columbus helped spread maize globally. As early as 1550 the hillsides of China were being cleared of trees to make way for this super-fast-growing crop that thrived on the sunny slopes of drier, higher terrain unsuitable for cultivating traditional varieties of rice. The population boomed. By the eighteenth century European farmers were increasingly turning to maize as a source of feed for their livestock and as a staple crop for human consumption. By the early twentieth century *polenta*, a dish made from boiled cornmeal, was firmly established as a premium food in the north Italian diet.

The conquest of maize in Europe had further political ramifications. Balkan peasants learned to grow the crop at high altitudes, sustaining clandestine alpine communities whose agricultural riches could be kept hidden from their Ottoman rulers. According to at least one authority, such practices *'weaned the future political independence of Greece, Serbia and Romania from mountain cradles. In this corner of Europe an American product really did nurture freedom.'*

Nineteenth-century Europeans and Spanish colonists who grew too dependent on this wonder crop became its victims just as much as those who were sacrificed by the customs of the ancient Americas. They suffered from pellagra, a vitamin deficiency disease that causes chronic insomnia, skin disorders, dementia and extreme sensitivity to light. The ancients of America added lime to their crops to avoid this condition but Europeans knew of no such fix. People in Spain, Italy and the southern United States who were almost totally dependent on maize were especially badly affected. Many people suffered and thousands died. Victims were kept in the dark, their disfigured faces inculcating a fear of vampires. The disease lasted until the mid-1920s when, finally, Western scientists rediscovered the cure – a simple dietary supplement to boost levels of vitamin B_3 (niacin).

Today maize is at the centre of yet another controversy. The USA cultivates more 'corn' than anywhere else in the world – about 270 million tonnes annually – over a quarter of the world's total. While much of it goes into feeding livestock, with all the consequences for global carbon emissions (see page 231), a great deal also goes into making cheap sweeteners for use in popular modern foods.

High-fructose corn syrup (HFCS), first synthesized in 1970, is produced using processed maize and is now America's most common sweetener, found in everything from fizzy drinks to cookies. Since its introduction, obesity in children aged between four and nineteen – the major consumers of these goods – has risen from 4 to 15.3 per cent. With more than six million US children now classified as clinically overweight, obesity, which leads to a greater risk of type-two diabetes, is now America's greatest health problem. Recent reports suggest that corn syrup is a '*big part of the problem*'.

The race to wean modern economies off their reliance on fossil fuels is now propelling the human–maize relationship in yet another direction. The production of 'carbon-neutral' biofuels, in a process that often involves the intensive cultivation of maize, is being promoted as a possible solution to global warming (see also algae, page 44; and yeast, page 302). This is because the process releases only as much carbon as it removes from the atmosphere. Unlike the burning of fossil fuels, there is no net increase. But adapting maize for this new purpose comes at a considerable political and economic price. In 2007 global food costs soared partly as a result of subsidies given to farmers to grow maize not as food but as a feed stock for fermentation to produce ethanol for fuel. The ancient and historic association between politics and *Zea mays* looks set to continue into the unforeseeable future.

10

On Material Wealth

*How certain species were
perfectly suited to providing
the wealth necessary to make life
comfortable for civilized man.*

Horse

FAMILY: EQUIDAE
SPECIES: EQUUS CABALLUS
RANK: 58

*The hard-working, quick-learning servant of humanity –
in war and peace*

THE PROPHET MOHAMMED is said to have ordered a herd of horses to be left without water for seven days. When they were released they raced towards an oasis, but before they got there the prophet sounded his war horn, calling the horses to battle. Five mares ignored the water and answered the call. Legend has it that all pure-blood Arabian horses descend from these five obedient mares.

The history of humanity is inextricably tied to the domestication of the horse. Today's pony clubs, flat-racing derbys or even Olympic eventing tournaments do not in the least reflect the historic impact of horses on man – and vice versa. Horses rose to pre-eminence among animals that learned to live with people because they provided a means of rapid transport and a source of labour in fields.

The family Equidae, to which modern horses, zebras and donkeys belong, arose about forty million years ago in what is now North America. They evolved from a fox-sized woodland-dwelling ancestor called *Hyracotherium*. As the Earth's climate cooled and grasslands replaced woodland, the need to flee from predators such as lions and sabre-toothed cats in the more exposed landscape meant that the fastest runners with the longest legs had a better chance of survival. These characteristics were passed down through the generations. By about five million years ago, after a series of evolutionary twists and turns, a wild animal had evolved called *Plesippus*, which looked much like a small modern-day horse. At some point during the ice ages (which began *c.* 2.5 million years ago) this species crossed the Alaskan land bridge into Asia, where it made its home on the extensive Eurasian steppes.

Here, above and around the Black Sea, the fate of wild horses and people began to merge with dramatic effect from about 6,000 years ago. Until then horses were probably killed by human hunter-gatherers for their meat. Taming the instinct of fear out of these creatures may have come about as an incidental effect of rearing foals as a source of meat. By about 3500 BC, however, there is evidence of nomadic people experimenting with domesticating mature horses, still mostly the size of ponies, and discovering that with skill and training these animals were more useful alive than dead.

The Botai culture, Neolithic people living in the region now called Kazakhstan, may have been among the first to tame wild horses. Evidence from the remains of horse teeth studied by researchers at Exeter University, England suggest riding bits and harnesses were in use as long as 5,500 years ago. They have also found the remains of food and drink in pottery vessels that suggest horse milk may have been consumed and even fermented into an alcoholic drink known as *koumiss*. It is known for certain that by 2600 BC a hybrid breed of domesticated horse was being used in the Sumerian civilization of the Middle East because it is shown on the Royal Standard of Ur, a painted box unearthed by British archaeologist Leonard Woolley in the 1920s. The animals are depicted pulling chariots in which soldiers stand on a platform ready to fire arrows at their enemies.

The period between 2100 and 700 BC saw wave upon wave of men on chariots pulled by horses invading civilizations from Egypt (Hyksos invasions) to India. This brought huge cultural changes throughout Europe, Africa and Asia. Societies that made a living through trade (such as the Minoan and Indus valley civilizations) and which worshipped a feminine deity of agricultural fertility (the

Mother Goddess) were from about 2000 BC displaced by violent horse-mounted cultures that originated in the grassland steppes near the Black Sea. The Minoans, with their love of animals, as depicted on their frescoes (showing images such as leaping dolphins), were replaced by Mycenaean invaders, whose tombs were full of objects made for war. The egalitarian Indus valley society of trades, crafts and commerce disappeared, to be replaced by chariot-riding invaders as described in the Vedic texts – probably dating from around 1700 BC. Class distinctions, initially into four castes, put soldiers and priests at the top of society with farmers and 'untouchables' at the bottom.

In Egypt the shock of Hyksos charioteers sweeping down from the north and invading the lower Nile (c. 1674 BC) caused a major rethink of imperial strategy when it came to burying their human gods, the pharaohs. Ostentatious pyramids that advertised the presence of buried treasure were replaced by underground chambers hidden to protect them from looting and theft. In Europe communal burial in barrows was usurped by a new trend in which individual graves were built for victorious invaders, who were frequently interred alongside their weapons of war (including horses). The birth of a new social hierarchy also meant settlements were built on the top of hills for protection and defence, not in lush valleys where the soil was most fertile.

All this happened because of the military power provided by the horse. By 700 BC the union of human and horse was even more closely melded. Selective breeding had increased the size of animals so that it was no longer necessary to wage battle in an unstable, hard-to-manoeuvre chariot with its highly vulnerable charioteer. Nomads who herded flocks of animals (sheep, cows, goats) used their horses as a complete life-support system. They could literally live (even sleep) in the saddle. Horses also carried their rations, and sometimes these nomads even managed to ferment milk when on the move using leather pouches where it would curdle into cheese (to avoid the ill effects of milk intolerance). Horsemeat from old animals was carried under the saddle as jerky. In extreme conditions they would even make a cut in their horse's neck and drink its blood as a source of protein (see also mosquito, page 130).

The history of Asia and Europe would have been utterly different were it not for a series of mounted nomadic invasions – probably provoked by climate change and population growth. The Huns, led by their famous leader Attila, made mincemeat of the declining Roman empire in the fifth century AD, following which central European political control collapsed completely for several hundred years.

Mounted cavalry remained the mainstay of human conflict until well into the twentieth century. Muslim horsemen, experts in the art of selecting and breeding light, manoeuvrable and highly obedient breeds (following the example set by Mohammed) charged out of the Middle East in the seventh century, conquering settled civilizations across Asia, Africa and Europe. Speed, reconnaissance and the precious element of surprise were theirs thanks to warfare conducted on horseback. Mongol chieftain Genghis Khan, born 1162, marshalled the most lethal nomadic force of all time, and his successors built the world's largest contiguous empire from China to Hungary. Mongol children were taught (by their mothers) to ride on horseback by the age of three, and by the age of five they could shoot with a bow while riding. Such complete symbiosis with the horse was the secret to their military success and, at least until the widespread use of gunpowder, settled societies had little or no defence.

Even after the arrival of infantry defences in the form of muskets, rifles and revolvers (such as during the American Civil War, 1861–5), the cavalry charge remained the most potent military tactic. Meanwhile it was the pulling power of horses that made possible the deployment of large cannon, used extensively by European armies from the seventeenth century. Barbed wire, trenches and machine guns eventually broke the horse's grip on the killing fields of World

War I (where an estimated eight million horses died), although by this time conditions were no less treacherous for infantry than for cavalry. It was the emergence of motorized armoured transport – tanks and jeeps – in World War II that finally took horses off the battlefield, although they remained much in evidence for hauling ammunition and supplies. Since then horses have continued to be used by police forces for crowd control and remain at the heart of ceremonial military traditions in many countries.

Horses came to have such a profound impact on the history of human civilization for two main reasons. The first concerns their genetic versatility. Mohammed's instincts about the benefit of breeding steeds to suit particular human needs have been consistently and successfully pursued at every stage of the human–horse relationship. Small feral ponies were 'domesticated' into species that could support warriors in battle. Those that survived the rigours of life with nomadic tribes were inevitably those that were best suited to the task. By the time of the Mongol invasions, steppe nomads had the most vigorous, robust, obedient and flexible breeds in the world.

The other reason why horses have had such a major impact is their highly sensitive nature. They can instantly tell when a fly lands on their back. Their ability to feel even the slightest movement of a rider is what enabled Mongol horsemen to control their horses hands-free. Without such sensitivity it would have been impossible for these raiders to fire volleys of arrows while charging past their enemies at thirty miles per hour. Recent research indicates that horses are naturally highly social creatures with complex group dynamics that moderate the aggressive characteristics of any one individual, providing social cohesion. It is this rational flexibility that has allowed horses to master their instincts and tolerate the most unnatural and terrifying environments, from battlefields to noisy urban streets, all while trying to follow the commands of a highly demanding bipedal ape: man. Given that horses are prey animals, whose instinct is to flee from threats and danger, their 6,000-year relationship with man shows an extraordinary level of adaptability and an exceptional capacity for learning and understanding.

The upshot is that today horses are another of the modern world's most numerous species. China has an estimated eight million, Mexico 6.2 million, Brazil 5.9 million and the US 5.3 million. Compare this to its closest wild relative, the zebra, whose worldwide population numbers only a few hundred thousand with several varieties on the verge of extinction.

Camel

> **FAMILY: CAMELIDAE**
> **SPECIES: CAMELUS DROMEDARIUS**
> **RANK: 87**

A superbly adapted pack animal that made merchants rich
through the trade of gold, spices, salt and slaves

MANSA MUSA WAS a medieval west African emperor who ventured across the Sahara Desert on a once-in-a-lifetime pilgrimage to Mecca. He set out in 1324, accompanied by a retinue of 60,000 subjects and 12,000 slaves, stopping off at Cairo for a spot of retail therapy. So memorable was his visit that it was still being discussed twelve years later, according to a contemporary Arab historian.

Musa's wealth came from a mountainous region of west Africa called Bambouk, situated in the heart of his empire of Mali. At that time the region's three gold mines supplied more than half of all the bullion traded in North Africa, Europe and Asia. Such riches were entirely dependent on being transported overland across the inhospitable desert to the Mediterranean markets of the Arab and Christian world. Without Musa's caravan of eight camels, which between them carried more than two tonnes of gold, medieval history's most celebrated shopping expedition could never have taken place.

Camels, like horses (see page 246), evolved in North America about forty million years ago. The arrival of human hunters quickly led to their extinction *c.* 12,000 years ago (see pages 191–196). Fortunately, by then, through various migrations across the Alaskan land bridge, stocks had already spread throughout Asia, reaching as far as the Middle East. Two-humped varieties (called Bactrian camels) thrived in the wetter climate of central Asia, whereas one-humped dromedaries, now by far the most successful species, lived in dry hot Arabian scrublands – habitats that became increasingly common throughout the Middle East as the climate dried out from *c.* 5000 BC.

Thanks to a range of brilliant evolutionary adaptations, camels can survive in places other mammals fear to tread. Contrary to popular belief, they do not store water in their humps. These are fat deposits that help camels survive long periods without feeding. Humps also help conserve water by keeping the heat-insulating fat supplies away from the bulk of their bodies. Camels can also tolerate a wide range of body temperatures (from thirty-four to forty-one degrees Celsius) without breaking into a sweat. Other water-saving survival techniques include long legs that keep their bodies as far as possible from the scorching hot desert ground, oval-shaped blood cells that maintain a regular blood flow even when dehydrated, and nostrils that collect and return water to their bloodstream from vapour exhaled in their breath.

It is one of human history's most fortunate accidents that creatures well enough adapted to trans-desert migrations also happened to tolerate living in close proximity to people. Tribes living on the coasts of Yemen about 4,000 years ago are thought to have been the first people to domesticate the wild camels that roamed in the deserts that began just a few miles inland. These tribes lived on seafood and regularly sailed across the Arabian Sea to the east coast of Africa (present-day Somalia). The local manufacturing of incense gave them a reason to travel overland to markets in the Near East. Crossing the Arabian Desert with goods to take to market would have proved impossible without camels as pack animals for carrying water. Domestication was probably assisted by the lack of natural camel predators, making these animals naturally less fearful of humans. Salty seafood, such as sardines, proved to be perfect bait to lure the camels close.

By 2,000 years ago the incense route had given birth to a number of wealthy trading cities, including Petra (now in Jordan) and Mecca (in Saudi Arabia). Camels became the chief means of transport across the Middle East, establishing overland trade routes connecting the Mediterranean to southern and eastern markets across the Syrian and Arabian deserts. They were also exported from Arabia by sea to Somalia, where camel-breeding thrived on the arid East African terrain, eventually providing Berber traders in North Africa with a powerful pack animal with which to transport their own commodities in the form of salt, slaves and gold across the expansive Sahara Desert.

Following the seventh- and eighth-century Islamic conquests of the Middle East and North Africa, Muslim traders had a huge tactical advantage over Christian Europe with their control of overland trade routes made usable thanks to the resilience of camels. Such networks connected the Mediterranean with the gold hub of Sub-Saharan Africa and the silk and spice markets of eastern Asia. Between c. AD 800 and 1450 the world's richest trading cities were the Muslim capitals of Cordoba, Cairo, Baghdad and Damascus, all of which thrived on camel power. The old quarters of these cities provide a reminder of the advantages of four legs over wheels – their narrow, winding streets were not designed to accommodate a horse and cart.

Mansa Musa's ostentatious expedition through Cairo didn't go unnoticed by European merchants. Indeed it was as a result of his visit that map-makers began to include his empire on their charts, later used by Portuguese maritime navigators to find ways of circumventing the camel traders' overland monopoly. As a result, from c. 1450 they established their own overseas trade routes, cutting out the desert caravans of the Sahara and their Muslim middlemen.

The camel remained vital for maintaining cross-desert links

throughout the world right up until the advent of motorized vehicles in the 1930s. Camels, along with their Afghan handlers, were even transported and deployed in the outback of Australia from the 1860s, amply proving their value to European settlers eager to trade gold and other raw materials from mines buried deep in that continent's arid interior. There was even an experiment to introduce camels to America at about the time of the Californian gold rush (*c.* 1848–55). This was restricted, however, for fear of equipping Native Americans with such a versatile, economically useful animal.

Yet camels, despite their immense value as a wheel-beating transport system, were never quite as globally successful as the domesticated horse (see page 246). The most highly prized domesticated animals in history are those that humans find equally effective in war, and camels were far too unstable to be reliable platforms from which to unleash a volley of arrows. Although the creatures could run at a constant twenty-five miles per hour, camel cavalry delivered nothing like the psychological impact of the sound and sight of horses charging at full gallop. Camels, for all their brilliant adaptations to desert conditions, simply aren't manoeuvrable enough to compete with horses in battle.

The internal combustion engine finally revived the wheel's fortunes in places where, for as long as 3,000 years, domestic camels had kept it in the shade. Camel herders still survive, however, especially in Somalia and Ethiopia, which between them host as many as half of the world's fourteen million population. While two-humped varieties are in decline (1.4 million), in Australia wild populations of one-humped camels – descended from those that escaped their handlers – are thought to number as many as 700,000 in the central outback. With their populations growing at an impressive rate of 11 per cent per year, it seems that, like pigs (see page 221), these are creatures that can readily re-adapt to life in the wild.

Perhaps one day, if and when the era of motorized transport comes to an end, populations of camels like these will resume their former role as essential pack animals. Just think of all the advantages: no need to build or maintain roads, breakdowns virtually unheard of and 'vehicles' automatically programmed to find their own fuel even in the harshest environments . . . that takes some beating!

Cotton

FAMILY: MALVACEAE
SPECIES: GOSSYPIUM HIRSUTUM
RANK: 39

An environmentally demanding natural fibre that provoked a modern contest for industrial and capital supremacy.

COTTON'S MODERN CONQUEST began in the early seventeenth century with the triumphant colonization of the Americas by smallpox-bearing Europeans (see page 15). Desperate to find a crop that could be traded to make European colonization economically viable, New World pioneers experimented with planting cotton – along with tobacco (see page 297) and sugarcane (see page 206). In 1607 the first cotton seeds were sown by colonists along the James river in Virginia. This crop soon became the biggest generator of economic wealth for what were to become the southern United States.

But turning raw cotton cloth into finished products was strictly banned in the British colonies. Imperial officials preferred to keep the profits to be made from manufacturing finished goods (such as clothes) close to home, where additional supplies of raw cotton were also beginning to arrive from trade with India. Potential fortunes were on offer if only a way could be found of cheaply converting raw cotton into material wealth – bales of fibrous cellulose into mass-produced sheets, shirts and trousers. The business of automating the processes of spinning, weaving and dyeing was therefore reliant on British entrepreneurs, whose rational problem-solving skills spun into overdrive at the prospect of turning cotton's short white fibres into yarns of strong, long material that could easily and cheaply be woven into cloth.

Such commercial incentives meant that between 1738 and 1800 a slew of ingenious human inventions transformed man's ability to harness the wealth of cotton. By 1788 more than 210 cotton mills straddled the fast-flowing rivers of Lancashire. The city of Manchester – dubbed 'Cottonopolis' – lay at the heart of this industrial revolution, leading to the creation of enormous wealth for British companies that converted cotton grown in colonies such as America and India into clothes for sale (often back to the very same colonies). Cotton became the economic bobbin around which the systems of today's globalized trade were originally wound. Japan's industrial transformation was similarly triggered by cotton. Following the enforced opening of its economy to global trade from 1853, the industrialization of Lancashire was transplanted almost piece by piece to the prefecture of Osaka in Japan. Bales of raw cotton imported from India were processed in Japan, initially using machines supplied by Britain, to produce finished cotton goods for export to China and back to India.

As with other crops that appeal to humans (such as sugarcane, see page 206, and rice, see page 236), harvesting cotton was an extremely labour-intensive affair. In the profit-hungry environment of early modern European colonization, its rise to economic predominance made sense only if the crop could be gathered cost-effectively. As a result, there was a dramatic increase in the use of slaves. The grim harvesting of cotton's material wealth was paid for by millions of kidnapped African men and women traded by European merchants for finished cotton goods, ammunition and arms. Inventions such as Eli Whitney's cotton gin (1793) made it possible to clean cotton fifty times more quickly by using horse power. The human cost of such scientific progress was that demand for raw cotton soared. By 1803 more than 20,000 African slaves were being imported annually and sold to work on American cotton plantations. By 1850 the United

States was producing as many as three million bales of raw cotton, hand-picked by African slaves, for export to Britain each year.

Injustices in India were no less profound. The effect of pitiful wages earned by cotton-picking labourers eventually led to a groundswell of nationalistic indignation, personified by their independence leader, Mahatma Gandhi. It was, said Gandhi, a scandal that raw cotton was exported for processing in the UK and then sold as finished goods to the Indian nobility. Meanwhile, profits from the transportation, manufacturing and sale of cotton goods grown in India were bagged by British-owned enterprises. The 'free-trade' dynamics of such inequality, rooted in the growing and processing of cotton, were instrumental in India finally gaining its independence from Britain in 1947.

Cotton's most recent legacy, at least in terms of its relationship with modern human civilization, is ecological. This shrub doesn't yield its magic fibre for nothing. Unlike legumes, it has no self-fertilizing symbiosis with bacteria such as *Rhizobia* to nourish soil (see page 30). Instead, cotton-producing *Gossypium* is heavily dependent on plentiful supplies of fertilizer and water while its susceptibility to infestations and diseases such as boll weevil is notorious. To satisfy human demand, therefore, cotton has become the most human-dependent crop of all time. Pesticides, artificial fertilizers and massive irrigation projects have poisoned ecosystems and emptied seas. Shortly before the collapse of the Soviet Union in 1989, the full horror of Russia's urge to become self-sufficient in growing and processing cotton (dubbed 'White Gold') was revealed to the world in a film called *Psy* ('Dogs') by Soviet director Dmitry Svetozarov.

Russia's cotton industry began in the 1870s but reached peak capacity in 1976. Its cotton plantations were irrigated by the diversion of two rivers that previously poured into the Aral Sea, once the world's fourth largest lake (68,000 square kilometres). Between 1960 and 1998 the amount of water taken to supply Russian cotton fields doubled, causing the volume of water in the lake to shrink by a staggering 80 per cent. What was once a vibrant marine ecosystem that supported 40,000 Russian fishermen is now little more than a dead brackish puddle. As much as 90 per cent of the former lake has become desert.

Cotton is therefore a crop synonymous with the industrialization and globalization of the modern world. Despite the growth in the use of artificial materials since the 1950s (such as polyester, nylon, acrylic), cotton is still the most ecologically demanding and environmentally damaging mass-produced natural fibre grown and processed all over the world today.

Rubber Tree

> **FAMILY: EUPHORBIACEAE**
> **SPECIES: HEVEA BRASILIENSIS**
> **RANK: 76**

*A relative late-comer to the crops that have
most influenced humanity*

BEGINNING IN ABOUT 1740, when European exploration of
the New World was still at its height, two Frenchmen set out on an
expedition down the Amazon. What most impressed Charles Marie
de la Condamine and François Fresneau as they explored the interior of Peru was how the natives living on the banks of the Amazon
seemed to 'milk' a certain type of tree. The thick white substance,
which the Frenchmen called 'latex' (derived from the Latin word
for 'milk') was processed in makeshift smokehouses, turning gooey
gum into elastic sheets that we would recognize today as a type of
rubber. The Frenchmen sent samples back to their home country,
where they set about conducting a series of experiments to see what
possible commercial uses this mysterious substance might have.

The native Amazonians were in fact pursuing a most ancient tradition – one established more than a thousand years before. A people
called the Olmecs (meaning 'rubber people'), who lived in Central
America, were the first civilization known to have used sap from a
tree called *Castilla elastica* to make rubber. They mixed the latex with
the juice of the morning glory plant (*Ipomoea alba*), dried it and cut
it into strips to construct bouncy rubber balls around a solid core.
The balls were used for playing a sacred game called *ulama*. Teams of
players would try to propel the ball through a hoop in a high stone

wall without using their hands or feet. Losers were often sacrificed to the gods, and their heads sometimes used to form the core of the next generation of sacred rubber balls.

The latex tapped by the Amazon natives that the eighteenth-century French explorers had observed came from a different species – the Para rubber tree, or *Hevea brasiliensis*. This material, the Frenchmen found, could be used rather successfully for waterproofing clothes by dipping fabrics in a solution of latex and turpentine. For the next hundred years small quantities of rubber were exported, mostly as a curiosity, from the Amazon to Europe and North America, where it was used, among other things, for making waterproof shoe coverings known as galoshes.

Rubber was not an instant hit in the rapidly commercializing and industrializing nations of Europe because it suffered from two serious flaws. When it was too cold it turned brittle, losing all its elasticity. When it was too warm it became soft and gooey. Attempts to find a variety that didn't go brittle when cold and sticky when hot were handicapped by the fact that a rubber tree starts producing milky sap only after six years of growth. It was therefore left to human curiosity and ingenuity to find ways to compensate for what natural genetics could not achieve in an acceptable human time-scale. It was found that the problems could be solved by heating latex laced with sulphur to 170 degrees Celsius for ten minutes. Two quite different entrepreneurs discovered this process – called vulcanization – between 1835 and 1843: the usually bankrupt maverick Charles Goodyear in America and a shrewd Protestant businessman called Thomas Hancock in Great Britain.

Vulcanization transformed global demand for rubber. The timing of this discovery was impeccable. High-pressure steam power, invented by Cornishman Richard Trevithick in 1801, was by now being used for self-propelled ships and railway trains to revolutionize transport all over the world. The elastic properties of vulcanized rubber could be applied to make seals for steam engines. The old method of sealing the steam cylinders, involving oil-soaked leather, was inefficient when subjected to high pressure, allowing steam to escape, reducing the engine's power and efficiency. Not so with vulcanized latex, which preserved its elasticity even at high temperatures, allowing steam cylinders to be fitted with seals that didn't leak. Vulcanized rubber seals led to ever more powerful engines, deployed in everything from cotton mills to electricity generating stations.

Rubber springs were also used to make the suspension for railway carriages, horse-drawn carriages and trams, making journeys more comfortable. With the invention of the bicycle and later the internal combustion engine in the nineteenth century, rubber found a new use in the manufacture of pneumatic tyres (commercialized by John Dunlop in 1885). By this time demand for rubber was truly enormous and the amount that could be supplied from wild trees tapped by natives in the Amazon basin simply wasn't enough, even though South American exports of rubber had rocketed from 200 tonnes a year in the 1830s to more than 20,000 tonnes by 1890.

Opportunities to cash in on this commodity boom were not missed by imperial wannabes such as Leopold II, King of Belgium (ruled 1865–1909). Leopold established his own rubber-growing empire in what is now the Democratic Republic of Congo. Latex milked from this region had traditionally been used by natives to attach heads to their hunting spears, but unlike latex from the Amazon it came from a wild vine called *Landolphia*. Leopold's private army used forced labour to extract rubber worth an estimated £14 million between 1898 and 1905. The price paid by the natives was horrific: many had their hands and penises cut off when they failed to deliver sufficient quantities of rubber. The population of the Belgian Congo dropped by ten million during Leopold's years of tyranny. Some died from a concurrent sleeping sickness epidemic (see tsetse fly, page 138) but most deaths were due to unimaginable levels of brutality. British clerk Edward Morel and civil servant Sir Roger Casement finally exposed the genocide, which Leopold had disguised under the auspices of a humanitarian project, in 1906.

Growing international demand for vulcanized rubber in the last quarter of the nineteenth century propelled Kew botanical gardens in London, internationally renowned as one of the world's leading centres for horticultural excellence, to employ the services of a struggling freelance British explorer, Henry Wickham. In 1875 Sir Joseph Hooker, then director of Kew, offered Wickham a reward of £10 for every 1,000 rubber seeds that he could collect from the Amazon basin and have transported back to London. With unbridled enthusiasm (and plenty of help from bribed natives) Wickham extracted as many as 70,000 seeds and shipped them back to London via Liverpool, earning himself a handsome £700 (and later a knighthood) for his troubles. When the Brazilian authorities discovered what had happened they branded the Londoner a despicable thief for carrying out an 'exploit hardly defensible in international law'. But, strictly speaking, at that time there was no official prohibition on the export of seeds.

Wickham's *Hevea brasiliensis* seeds arrived at Kew in June 1876 (see plate 6, bottom). About 10 per cent germinated in the botanical garden's greenhouses, of which almost 2,000 were immediately exported to Ceylon, now Sri Lanka. The plan was to establish rival rubber plantations under British control some 10,000 miles away from their native country of origin. After several unsuccessful cultivation attempts (and further shipments of seedlings from Kew), twenty-two healthy plants were eventually forwarded from Ceylon to the botanical gardens in the British colony of Singapore. Sir Henry Ridley, director of the gardens, was under no illusion as to the potential value of rubber plantations. Under his stewardship the seedlings were sent to Malaya, where an initial 2,000-acre plantation was established in 1898. By 1920 colonial Malaya's newly introduced rubber trees covered more than two million acres of what was once tropical rainforest and supplied the bulk of the British empire's ever-increasing appetite for vulcanized rubber.

Demand for rubber continued to rise, mostly as a result of the popularity of the motorcar from the 1930s onwards. Between the two world wars scientists set out to find a way of making rubber artificially, using oil as a raw material. Synthetic rubber gradually began to supplement natural rubber, especially when sources of natural latex from south-east Asia were interrupted during World War II (1939–45). Today about 40 per cent of the world's annual output of sixteen million tonnes of rubber comes from milking nature's prime latex-producing species, *H. brasiliensis*. The rest is synthesized from oil.

Silkworm

FAMILY: BOMBYCIDAE
SPECIES: BOMBYX MORI
RANK: 86

A highly domesticated moth that has enriched
civilizations for thousands of years

QUITE A FEW invertebrate animals produce silk as a natural product. It forms a vital part of their survival strategies in the wild. Spiders spin webs of silk to catch flies, and bees sometimes produce a silk-like substance to help construct their nests. But the sort of silk that has captured the imagination of men and women for at least the last 6,000 years belongs to a type of moth, *Bombyx mori*, whose caterpillar larvae feed on the leaves of mulberry trees.

Legend has it that Leizu, wife of an ancient Chinese emperor, was the first person to discover the potential wealth locked up in the cocoon of *B. mori*. She had been charged by her husband to find out why his grove of mulberry trees was looking so sick. As she took a break from her investigations a cocoon from a caterpillar dropped into her cup of tea. As she looked down she saw it unravel. Entrepreneurial curiosity took over and, it is said, Leizu worked out how to weave the material into fabric.

As Leizu discovered, when dipped in hot water, fibre from the cocoon of *B. mori* unravels into a single thread, sometimes stretching up to *one kilometre* long. This makes silk quite unlike wool (see page 224) or cotton (see page 257), which yield only short fibres that stretch to a few centimetres at most. They have to go through a complicated process of spinning to turn them into long threads of useable yarn. Making fabric out of silk is simpler. First, kill the caterpillar by dipping the cocoon in boiling water (to avoid it hatching and therefore severing the thread) and second, twist and wind the long fibres from several cocoons on to a single bobbin to make a robust, highly versatile material, ideal for weaving into cloth.

Silk is also one of the toughest naturally occurring substances in the world, a legacy of its original evolutionary purpose as a sturdy changing room. It didn't take long, once the secret of silk making had been discovered, for human reason to put such a valuable material to work. Some of the earliest battle garments in history were tightly woven Chinese silk vests, which offered a degree of protection against incoming arrows. Genghis Khan (1162–1227) issued them to his Mongolian horsemen. Even in the late 1800s American bullet-proof vests made out of silk provided effective protection against shot from black-powder handguns.

But silk's biggest appeal was to the human senses. Thanks to the triangular arrangement of its molecules, silk reflects the light in a way that gives it an iridescent sheen. This, together with the fact that it is lightweight, warm and elastic (which stops it creasing), means silk is ideal for clothing. Finally, when dyed, cloth made from silk has a deep, rich appearance that makes those who wear it look and feel really rather special. This is why kings, emperors, popes and aristocrats have traditionally chosen silk as their preferred material for clothing. Archaeological evidence suggests that people living along the Yellow river valley in China between *c.* 6000 BC and *c.* 3000 BC were the first to practise sericulture. Silk was so highly revered by ancient Chinese society that sumptuary laws were passed to prevent peasants from wearing the fabric.

Silk was used as a symbol of wealth, status and privilege in human societies – from Beijing to Cairo – right from its very earliest manufacture. Fragments of a silk thread recovered from the hair of an Egyptian mummy (*c.* 1070 BC) indicate that some of the earliest-known trade contacts between east Asia and the Mediterranean were based on Chinese silk. Silk garments became so popular in Roman high society that a decree was passed by Emperor Tiberius (42 BC– AD 37) in an attempt to prevent so much gold passing into Chinese hands. When the Byzantine Greeks took over the eastern part of the Roman empire beginning in the fifth century AD, silk was already an essential part of the imperial arsenal for re-establishing law and order. Justinian I (emperor of the eastern Roman empire AD 527–65) had a three-pronged approach to restoring imperial authority: military reconquest, legal renaissance and the establishment of indigenous silk production.

To Justinian, the priority was to set up his own silk-making industry. He sent two Nestorian monks as spies to China, from where they successfully managed to smuggle a batch of *B. mori* eggs back with them to Constantinople (safely tucked inside a bamboo cane). Justinian established mulberry groves and imperial silk factories in Greek cities such as Corinth and Thebes. Tribal leaders and ambassadors from other cultures were treated to displays of royal opulence inside the capital Constantinople, and given sumptuous gifts of silk in exchange for political alliances. This proved that societies can survive – at least for a while – by diplomatic cunning as much as military might. However, such opulence eventually contributed to the empire's demise at the hands of Christian crusaders. They sacked the imperial city in 1204 when they were supposed to be ousting Muslim 'infidels' from Jerusalem. As a result of the raid, *B. mori* and its human cultivators spread farther eastwards into Christian Europe via Norman Sicily.

European lust for the richly coloured shimmering fabric reached its zenith in France in the fifteenth century. King Louis XI (reigned 1461–83) applied the same tactics as Justinian, although nearly 1,000 years later. In 1480 he relocated an entire silk-growing and processing industry to the French city of Tours in an effort to avoid paying Italian merchants exorbitant prices and to reduce his country's spiralling trade deficit. The policy was reinforced under

On Material Wealth

King Francis I (reigned 1515–47), who granted the city of Lyon a monopoly on silk production. By 1560 more than 12,000 people were employed in this single city's silk-weaving industry.

It was here that Joseph-Marie Jacquard (1752–1834) invented a new type of loom, which could weave portraits out of silk cloth as if they were painted pictures. It worked by employing a series of punched cards (rather like a pianola scroll) that could control every individual line in the weaving of a single piece of cloth. Like woodblock printing, once the cards had been designed, they could be used again and again. Five years after the death of Jacquard this mechanized form of silk weaving inspired British inventor Charles Babbage to use punch cards to programme his famous 'Calculating Machine' – the world's first programmable computer.

Another country that profited from wealth generated from silk was Iran. Located at the heart of the ancient overland trade routes that connected Asia to Europe, a succession of Persian regimes had long been successful at adding value to raw silk from China by turning the shimmering fabric into luxurious textiles. The Safavid dynasty (1501–1722) was able to use the wealth generated by selling silk rugs to European powers, via merchants in Christian Armenia, to re-equip and retrain its military forces with firearms, which were beginning to replace traditional weapons. The Shi'a form of Islam is still predominant in that region of the Middle East today, largely as a result of the Safavids' silken success.

But *B. mori* caterpillars never quite reached the status of a globally dominant species, even if the silk products of their labours did. British monarch James I tried to grow 100,000 mulberry trees near his palace at Hampton Court, London, with a view to copying the success of the Byzantines and French, but for some reason his caterpillars never took to their newly adopted home, although limited production was successful at a castle in Kent called Lullingstone, and continued there until the middle of the twentieth century. The caterpillars weren't a big hit with colonialists in North America either.

Small-scale production succeeded in New Jersey in the 1800s but by then cheap Japanese silk imports precluded mass production in the United States. By the middle of the twentieth century the break in supply of raw silk as a result of World War II led to the creation of artificial alternatives. Nylon, first synthesized in 1935, caught on as a replacement for silk and quickly came to be used in everything from parachutes to ladies' stockings. The same plastic fibre has since replaced other natural fibres such as wool and cotton, especially for use in sheets, clothes and other consumer fabrics.

B. mori is probably the least fortunate of all the main domesticated species. After thousands of years of human selection these moths are now genetically so far removed from life in the wild that even as adults (when they are allowed to hatch out of their cocoons in order to lay fresh eggs) they cannot fly or feed without human assistance. Meanwhile, the vast majority are killed before they are allowed to mature in order to protect their 1,000-metre-long thread.

Eucalyptus

> FAMILY: MYRTACEAE
> SPECIES: EUCALYPTUS GLOBULUS
> RANK: 16

A highly invasive species uncorked by European colonists who discovered its treasures down under

EUCALYPTUS TREES EVOLVED between fifty million and thirty-five million years ago. Climate change and the geology of Australia largely explain why these angiosperms behave as they do today. Australia was once a part of Gondwana, the southern part of the super-continent Pangaea that began to break up *c.* 250 million years ago. Having avoided all other landmasses since, Australia has some of the oldest, least-disrupted terrain on Earth. In the absence of a good geological stir, typically provided by volcanic activity from colliding tectonic plates, life-supporting minerals and nutrients have been sucked dry. Species that have thrived are therefore those most intensely competitive for access to the scarce raw materials necessary for life. Over the last thirty-five million years Australia's climate has also become increasingly arid. Divining for water is therefore another essential survival skill that has made the critical difference between survival and extinction. Just as when mammals developed their own bag of essential survival adaptations in the shadow of the dinosaurs, so eucalyptus learned how to thrive in the dry nutrient-poor landscape of what is now modern Australia.

Their range of survival skills is breathtaking. First, they grow extremely fast – reaching full maturity in just six years. In a competitive environment, the quicker a tree can grow and establish its canopy to snatch as much sunlight as possible, the better (see also *Lepidodendron*, page 82). Second, eucalyptus trees have superb skills for finding water even when sources are buried deep in the ground. Long roots and rich, intricate connections with mycorrhizal fungi (see *Prototaxites*, page 76) combine to provide a powerful solution. But it is eucalyptus oil – famous today for its disinfectant, antibacterial, antifungal and anti-insect properties – that comprises this tree family's ultimate weapon. Rotting leaves and sticky gum (exuded from the roots and bark of the eucalyptus tree) kill off microbes and poison the surrounding soil, thereby warding off other species that might threaten supplies of vital nearby nutrients.

If a lightning strike happens to trigger a forest fire, the dense-growing eucalyptus gives the arboreal equivalent of a victory whoop. The whole forest goes up in flames all the more readily thanks to its highly volatile combustible oil. Then, just as a phoenix rises from the ashes, these die-hards grow back from charred stumps thanks to tubers and buds that lie buried under layers of protective bark. With all remaining competition removed at a sweep, the smouldering ash provides a perfect carpet of nutrients from which forests of eucalyptus regrow.

Aboriginal Australians co-existed with forests of eucalyptus for tens of thousands of years before the first European colonists arrived. Infusions made from eucalyptus oil were a traditional medicine used for treating bodily pain, sinus congestion, fever and colds. Oil from one of the most widely grown species today, *Eucalyptus globulus*, is still used to make cough sweets and as an antiseptic.

Potential for material gain was not lost on Captain James Cook (sailed 1768–79) and his crew when, at the end of the eighteenth century, they began to familiarize themselves with these highly successful species. Jospeh Banks (1743–1820), botanical adviser to King George III, was instrumental in ensuring eucalyptus seeds were brought back to Britain, where they were cultivated in botanical gardens such as Kew (see page 261). Within fifty years the prodigiously fast-growing properties of this tree family helped turn it into one of the most widely cultivated and globally transported groups of trees in human history. With demand for wood soaring as a result of high-pressure steam power, the building of railways and the construction of mines, species of eucalyptus (especially the red gum *Eucalyptus camaldulensis* and the blue gum *Eucalyptus globulus*) were, for ambitious European industrialists, like manna from heaven.

Wherever nineteenth-century European colonists went in their global pursuit of material wealth, eucalyptus trees followed in their wake. In 1850 the trees were introduced to California at the time of the gold rush to support the massively increased energy requirements of an immigrant population that soared by as many as 300,000 people in just three years. Eucalyptus trees respond well to human coppicing. They quickly recover after being hacked down thanks to their well-buried buds, developed in response to decimation by fire.

Edmundo Navarro de Andrade (1881–1941) was a Brazilian agronomist who has been credited with establishing more than *twenty-four million* eucalyptus trees in plantations that stretched alongside his country's rapidly expanding railway network. As a fast-growing hardwood it provided copious supplies of fuel for steam engines, sleepers for railway tracks, telegraph poles for communications systems and fencing to keep animals off the tracks. By 1925, de Andrade was exporting eucalyptus to the USA as experimental material for woodpulp in order to make high-quality brilliant white paper. Such pulp is among the most popular today, thanks to the advent of modern laser printers and photocopiers.

British authorities introduced eucalyptus trees into Kenya in 1902 to supply fuel for the Kenya-to-Uganda railway. In South Africa these same trees were introduced to provide props to support shafts in gold mines. Meanwhile, Spanish and Portuguese authorities authorized vast eucalyptus plantations as a source of revenue from woodpulp. They still dominate the forests of Extremadura on the western Iberian peninsula. The conquest of Asia was no less spectacular. China, Thailand and India all host enormous plantations established over the last 200 years to export pulp, firewood and satisfy the worldwide demand for the therapeutic properties of its oils, 70 per cent of which come from China.

But in the last twenty-five years, some countries have begun to realize that the Victorians' headlong rush into planting fast-growing alien species without first investigating the ecological side-effects on native species may not have been so wise. Since eucalyptus evolved in nature's toughest environments, when transplanted into wetter, more fertile, temperate zones, their ability to dominate all other life-forms turns them into agents of ecological destruction. Species of fast-growing evergreen eucalyptus make huge demands on local water resources. As a result, wetlands are frequently reduced to desiccated scrubland or desert. In South Africa species of Australian eucalyptus suck up more than 70 per cent more water than indigenous drought-resistant African acacias (see page 147). The result has been as catastrophic as that from cotton-growing around the Aral Sea (see page 259). Water flowing from river systems has reduced by three quarters. Such dry conditions also turn the habitat into a tinder box in which frequent forest fires benefit the well-adapted eucalyptus even more.

So acute have concerns become in some places that authorities have begun to turn the clock back by literally ripping these highly invasive trees out of the ground. Working for Water is a South African project launched in 1995 that is clearing as many as 200,000 hectares of eucalyptus plantations a year in a desperate bid to restore badly needed water supplies. Meanwhile, concern for wildlife has driven the Spanish authorities to declare the forests around Monfragüe near Trujillo a natural wildlife park in an attempt to protect some of Europe's rarest species of storks, vultures and eagles as well as the almost extinct Iberian lynx. The park has recently established a programme of systematically removing eucalyptus plantations and replacing them with cork and holm oaks to revive the local ecosystem and raise the water table. Oaks, unlike eucalyptus, are indigenous species whose life strategies have co-evolved with local wildlife including black feral pigs that feast on their prodigious quantities of acorns (see page 143).

How are we to deal with such a prolific success story as the eucalyptus in the environmentally conscious twenty-first century? Some see this family of trees as best placed to help absorb high atmospheric levels of CO_2 and form sustainable coppices for providing timber, paper and energy supplies. Others say they are at the sharp end of an ecocide spread by humanity's increasing population and lust for material wealth.

11

On Drugs

How certain plants and fungi diverted the development of human culture by bestowing a bevy of addictive habits.

Cacao

> **FAMILY: MALVACEAE**
> **SPECIES: THEOBROMA CACAO**
> **RANK: 81**

How the therapeutic charms of a South American tree
have been hijacked by the modern world's addiction
to high-speed confection

WHY ARE PEOPLE crazy about chocolate? Have humans always been so susceptible to its delicious flavour? What has been the impact of our apparent addiction, past and present, which stems from the seeds of this otherwise unspectacular South American tree?

When Carolus Linnaeus, the famous eighteenth-century Swedish botanist, first described this tree species, he chose its name especially well. *Theobroma* is the Latin for 'food of the gods' and *cacao* derives from *cacahuatl*, the name used by the Mayan and later Aztec people who cultivated these trees in what is now Mexico in the pre-Columbian era. Genetic evidence clearly shows, however, that the cacao tree did not originate in Mexico, which is where Europeans first encountered the hot, rich, aromatic drink made from its beans in 1519. Cacao trees are native to lowlands around the Amazon river in what is now Brazil and Peru. Once they have flowered, cacao trees produce fruits called pods. Each one typically contains thirty to fifty large almond-like seeds or beans, which are surrounded by a thick sweet pulp. The beans and the pulp are cut out of the pod and left in the sun, where they ferment naturally. Once dry, they can be processed into chocolate, either for drinking or eating.

The first archaeological evidence of human cacao cultivation dates from around 1100 BC. Jars from this era containing traces of chemicals from the processing of chocolate beans have been uncovered by archaeologists working in modern-day Honduras in Central America. They show that widespread cultivation and trade must therefore have existed well before this time since Honduras is a considerable distance from the trees' centre of origin in the Amazon lowlands. The indigenous peoples of South and Central America put a high value on the trade and consumption of chocolate. By the time of the late Aztec empire (c. 1440–1521), tribute payments from neighbouring states were made in the form of sacks of cocoa beans. When Hernán Cortés and his retinue of Spanish invaders first encountered Aztec Emperor Moctezuma their curiosity was aroused by this mysterious drink. According to their accounts Moctezuma drank fifty whipped cupfuls of hot chocolate from a golden goblet every day. His attending nobles also consumed copious quantities, they said, as many as 200 daily helpings between them.

Such high levels of individual consumption would, by most people's reckoning, amount to addiction. Additional evidence of chocolate's addictive qualities is borne out by the global effect cocoa beans have had, beginning with their cultivation, transportation and trade by Spanish-American colonists from the late sixteenth century onwards.

Naturally occurring active ingredients in chocolate include *theobromine* and *phenethylamine*. The former dilates the blood vessels, increasing the flow of blood throughout the body and brain, resulting in a feeling of general mental and physical wellbeing. The latter, which belongs to the amphetamine family of drugs, affects neural pathways in the brain, potentially leading to feelings of contentment and euphoria. It is the effect of this chemical that has given rise to cocoa's legendary aphrodisiac qualities. However, recent medical research suggests that chocolate's romantic power is probably derived more from tradition than from chemistry since in most people the liver metabolizes the chemical before it can have any significant impact on sexual arousal.

The initial spread of chocolate beans through Europe began innocently enough. Like Moctezuma, Spanish royalty and their attending nobles adopted the fashion, enjoying the flavour of drinking chocolate, but often adding vanilla (rather than chilli, preferred by indigenous Americans) to suit the European palate (see page 362). When the Spanish-groomed Habsburg Princess Anne of Austria married French King Louis XIII in 1615, she took the chocolate-drinking habit to France, where its aristocratic appeal spread further. For the following 200 years the therapeutic effects of chocolate drinking were therefore made available to upper-class people in the ancient regimes of early modern Europe.

But then, in the first half of the nineteenth century, all that changed. Aided by the forces of industrialization and mass-production in water-powered mills, chocolate's destiny took a dramatic turn. When Dutch chocolate-maker Conrad Johannes van Houten (1801–87) pioneered a pressing process for milling cocoa he removed the centre of the bean – called the nib – in order to make the resulting paste more digestible. This also had the commercial benefit of making chocolate paste easier to combine with other substances, such as milk and sugar.

As a direct consequence of van Houten's hydraulic press, chocolate could now be made into solid bars for immediate consumption, as opposed to concentrated cubes useful only for dissolving into a hot therapeutic drink. Milk-and-sugar-enriched solid chocolate bars had the natural bitterness removed (and with it most of the healthy chemicals), turning what was an exotic health drink into a sucrose-rich confection. Along with many other modern fast foods, mass-produced chocolate has substantially increased levels of tooth decay and played its own part in the acceleration of the modern pandemics of diabetes and obesity that plague the populations of developed countries today.

Once van Houten's patent expired in 1838, a clutch of enthused Victorian chocolate entrepreneurs leapt on to the chocolate-making bandwagon. Englishman John Fry brought out his first chocolate bar in 1847, followed by the Cadbury brothers two years later. As steam power began to take over from waterwheels, the cult of the modern chocolate factory, Willy Wonka-style, was born.

Unfortunately, Roald Dahl's story about the man who wanted to find an honest heir to his chocolate factory reveals a dark side to the worldwide spread of cacao since the emergence of chocolate bars in the 1850s. Child-like 'Oompa-Loompas' – Willy Wonka's mythical midget ethnic-looking factory staff that some say resemble African pygmies – still reflect the terrible human price paid today for mass-produced cocoa. Few consumers are aware of the living and working conditions in the world's largest cacao plantations, located in west Africa, as they gaze longingly at their Valentine's gift of a box of dark chocolates.

The Côte d'Ivoire (formerly known as the Ivory Coast) exports nearly half (46 per cent) of the world's current production of cocoa beans. Its plantations were established between 1890 and 1960 by French colonialists who sought to profit from their newly acquired territories by growing cash crops such as coffee (see page 283) and cacao. As a result, this war-torn nation is still dependent on whatever income it can gain from the export of its beans to chocolate-processing corporations in Switzerland, England and Holland via the London Commodity Exchange, which runs the international trade in cocoa beans. Families desperate to scrabble together a living from the harvesting of beans recruit every member who can help, however young. Other children are sold as slaves and trafficked into the Côte d'Ivoire from neighbouring countries such as Mali because the use of unpaid labour is the only way of keeping the prices demanded by international markets sufficiently low.

A US Department of State study published in 2005 makes dismal reading regarding the human impact of sustaining the modern global chocolate habit. In its report into human rights in the Côte d'Ivoire it said: '*Approximately 109,000 child labourers worked in hazardous conditions on cacao farms throughout the country in the worst forms of child labour.'*

What was once a natural product cultivated and consumed as a therapeutic remedy and royal delicacy by the indigenous people of central and southern America is now a mass-produced health hazard mostly grown by poverty-stricken African peasants, many of whom are children. The fruits of African labour feed the addictive habits of post-industrial nations made fat and diseased by chocolate.

Coffee

> FAMILY: RUBIACEAE
> SPECIES: COFFEA ARABICA
> RANK: 85

A sub-tropical plant that helped commercialize modern society with its stimulating effect on the body and mind

KALDI WAS A nomadic goatherd who lived in tenth-century Ethiopia. One day, it is said, his goats went crazy after munching the red berries of an unknown wild bush. Maybe out of curiosity, or perhaps boredom, Kaldi ate some of the fruits himself. Not long after, he was spotted dancing gaily among his goats by some monks who lived nearby.

The invigorating effect of coffee derives from caffeine, a naturally occurring chemical insecticide. *Coffea arabica* (and its other popularly grown cousin *Coffea canephora*) are plants that have especially high levels of caffeine in their seeds. The chemical also sours the soil, preventing other plant-life from growing too close and competing for resources. It is rather bizarre to think that the world's most popular human beverage is, in its natural state, a growth-stunting pesticide.

In humans caffeine is now known to inhibit the action of various receptors in the brain, resulting in a build-up of chemicals that induces a state of mental and physical stimulation that can last for several hours. When caffeine is taken regularly, the brain compensates by establishing more receptors, so that the stimulating effects of caffeine are reduced. However, if the drug is withdrawn, the excess number of receptors causes an imbalance in the behaviour-regulating systems of the mind, leading to headaches, nausea and irritability. These patterns of tolerance and withdrawal are classic symptoms associated with addictive drugs, to which the human brain seems particularly susceptible.

From Ethiopia, news of coffee's energizing effects spread to Yemen, on the southern tip of the Arabian peninsula, where trans-Arabian/African trade was already well established (see camel, page 252). Then it reached Muslims on their annual pilgrimage to Mecca, where the drink, which became known as the 'wine of Araby', was found also to stimulate conversation, trade and commerce. Coffee houses sprang up in Islamic trading cities, such as Damascus, Cairo, Baghdad and Istanbul. By the fifteenth century such establishments had become centres of trade and exchange between Islamic and European merchants. Coffee sharpened the wits of haggling traders to make sure each came away with the best possible deal.

European coffee houses brought the association between coffee and commerce even closer. Until *c.* 1650 European people depended on beer for liquid nourishment. In England the average family drank about three litres per person per day – children included – and businessmen traditionally consumed a pint during their mid-morning break. But by the middle of the seventeenth century, as coffee houses first began to appear in London, a dramatic transformation was under way. The stimulating, sobering effects of coffee fused with the entrepreneurial spirit of overseas exploration that swept across early modern Europe from the seventeenth century onwards. It was a powerful cocktail. By 1700, according to one estimate, there were a staggering 3,000 coffee houses in London alone – that's one cafe for every 200 people living in the city at that time. Drinking coffee became an addictive habit in the bourgeoisie, a fashion few social climbers, or those wanting to further their careers, could resist.

When Edward Lloyd opened his coffee house on London's Tower Street in 1688, little did he imagine that it would evolve into a vibrant meeting place for mariners, ship's captains, merchants and insurance brokers. By 1774 his humble cafe had transformed into the world-famous institution Lloyd's of London, which is still the largest insurance brokerage in the world. Meanwhile the posting of stock and commodity prices in Jonathan Miles' Coffee House, beginning in 1689, cemented the relationship between coffee and commerce even further. This cafe's stockbrokers eventually built their own premises in 1773, originally dubbed 'New Jonathan's', which later became known as the London Stock Exchange.

Journalism was another trade born in the coffee house. Political and literary topics were discussed with feverish enthusiasm in the coffee houses of London. As a result weekly newspapers such as *Tatler* (started in 1709) were founded to publish the gossip overheard (hence the name). The paper even used a coffee house as its editorial office. Meanwhile literary and philosophical giants of the French Enlightenment, such as Voltaire, Diderot and Rousseau, spent many of their waking hours talking and writing in the coffee houses of France.

By the beginning of the eighteenth century it was clear that the European appetite for coffee presented substantial commercial opportunities for overseas explorers and colonists – if only they could find places to cultivate the crop cheaply enough. Dutch traders were the first Europeans to establish coffee plantations overseas. Pieter van den Broeck (1586–1640), a cloth merchant, smuggled live coffee seedlings out of the Yemeni port of Mocha in 1616, despite the fact that their export was strictly prohibited by the authorities. He used the seedlings to establish the first plantations on Dutch colonial territories on the island of Java. Meanwhile, the popularity of coffee increased exponentially throughout Europe following the defeat of the Turkish forces outside the walls of Vienna in 1683. Polish soldiers found sacks of coffee beans in the camps abandoned by fleeing Turkish troops, seeding the start of a thriving coffee-house revolution among the liberated inhabitants of downtown Vienna.

Human deception, rivalry and cunning all played their part

in turning coffee into a global cash crop which today boasts an estimated annual trading value of $33 billion. The crop's biggest break came when sneaky French naval officer Gabriel DeClieu transported coffee to the Americas *c.* 1720 after taking an illicit cutting from a royal bush in the botanical garden belonging to King Louis XIV. After a stormy Atlantic crossing (in which he is said to have sacrificed his last water rations to keep the plant alive) DeClieu landed in the French Caribbean colony of Martinique. Originating from his cutting, plantations were soon established across the region, as well as on the South American mainland just north of Brazil in French-controlled Guiana. Towards the end of the eighteenth century, coffee from Brazil was re-introduced to Kenya and Tanzania, completing a global circumnavigation that had begun less than 1,000 years before.

Today coffee is by far the most popular chemical stimulant in the world, with an estimated *500 billion* cups drunk annually. To support such enormous consumption, 7.8 million tonnes of coffee beans are cultivated each year. Brazil accounts for about 40 per cent of worldwide supply. In that country alone the growing and harvesting of coffee employs as many as five million people, who tend to three million individual coffee plants.

Tea

> FAMILY: THEACEAE
> SPECIES: CAMELLIA SINENSIS
> RANK: 71

*Highly versatile leaves used for everything from
beverage-making to predicting the future*

FOR CHINESE PEOPLE, drinking tea is a habit as venerable as their 5,000-year-old civilization itself. According to an ancient legend, the trick of brewing the leaves of *Camellia sinensis* in boiling water was first discovered *c.* 2737 BC by the fabled emperor Shennong. One day he was drinking from a bowl of hot water when a sudden freak blast of wind blew a few leaves from a nearby bush right into his cup. As he looked down, he saw the colour of the water begin to change. Being a curious fellow Shennong took a sip. That's when he discovered the delightful and restorative effects of drinking tea.

Although our modern tea-drinking habit is regarded as an everyday comfort, for thousands of years tea was the most significant natural ingredient in Chinese herbal medicine and Asian spiritualism. For Laozi, a spiritual leader and philosopher who lived in the sixth century BC (at about the same time as the Buddha), tea was the elixir of life. On one of his journeys around China, during the violent Warring States period, Laozi is said to have been offered a cup of tea by a customs inspector named Yin Hsi. As they consumed the drink, conversation flowed. Captivated by what the philosopher said, Yin Hsi implored Laozi to write down his thoughts in a book so others could benefit from his wisdom. It is thanks to tea, therefore, that the *Dao De Jing* (The Way of Virtue), a collection of Laozi's most famous sayings, such as '*He who knows he has enough is rich*', has been passed down through history to become among the most influential writings in all Chinese culture.

It is known that tea cultivation was widespread in China by *c.* AD 800 thanks to a treatise on tea growing called *Cha Jing* (The Classic of Tea) written by a former circus clown called Lu Yu between AD 760 and 780. The book details all aspects of tea's origins, cultivation, processing, brewing, serving and drinking, showing that by this time the little leaves of *C. sinensis* had diffused deep into the heart of Chinese culture, from where the plant spread even farther to the east. Records suggest that tea plants reached Japan in the early ninth century, where they were used by priests during religious ceremonies. A Zen Buddhist monk called Eisai (AD 1141–1215) travelled extensively in China and brought green tea back to Japan. In 1211 he wrote a highly influential two-volume book called *Kissa Yojoki* (How to Stay Healthy by Drinking Tea) which promoted the tea-drinking habit enthusiastically. The book opens by saying: '*Tea is the ultimate mental and medical remedy which has the ability to make one's life more full and complete.*'

When Western mariners began to unlock the markets of Asia, beginning in the sixteenth century, tea was one of the first products they shipped to Europe. Sleek sailing ships called clippers formed the backbone of a maritime trade in which tea grown in China was transported back to markets in Amsterdam and London. But it wasn't the medicinal effects of tea, drunk in its unadulterated traditional Chinese form, that turned this tonic into the everyday classic we are addicted to today. The habit of sweetening tea and adding milk not only made the drink more suited to the sensitive European palate, but also gave a commercial boost to the sugar grown on European colonies on the other side of the world in the Caribbean and Americas (see page 206).

Between 1700 and 1950 sweet, hot, milky tea became Britain's favourite drink, transcending all barriers of class and gender. Anna Maria Russell, Duchess of Bedford (1783–1857), invented the Victorian aristocratic custom of taking afternoon tea, and later Joseph Lyons and Co. (established 1894) built a nationwide network of teashops and corner houses that sustained the spirits of Britain's middle- and working-class men and women throughout two world wars. From *c.* 1700 the habit of drinking tea spread to the Americas, even surviving the symbolic stigma of crates being tossed into Boston Harbour (1773) in protest at the importing of tax-free tea by British traders keen to undercut local American fare. So great was the demand for tea in Britain's domestic and overseas colonial markets that paying Chinese cultivators in precious silver bullion began seriously to undermine the empire's balance of payments. In the late eighteenth century directors of the Honourable British East India Company came up with two solutions to the problem.

The first involved the East India Company diverting some of its extensive farming resources in a region of India called Bengal into fields for growing poppies, from which they extracted opium. An illegal triangular trade was then established in which British tea traders would buy Chinese tea in Canton using credit notes that could be redeemed against opium smuggled across the Chinese border from India. In this way the British paid for tea using a drug grown by oppressed natives in India while China became addicted to opium (see poppy, page 321). The second concerned the smuggling of *C. sinensis* plants out of China to establish rival tea plantations in British-controlled India. Robert Fortune (1812–80), a gardener and orchid specialist with the Royal Horticultural Society, was tasked by directors of the East India Company with locating tea plants within China and sending seedlings back to India (a crime punishable by death). In 1848, disguised as a Chinese mandarin with shaven head and pigtail, the 35-year-old Fortune travelled to China under the pseudonym Sing-Wa (meaning Bright Flower). After travelling extensively up the Yangtze river and into the tea-growing

Wuyi Mountains of south-west China, he located and transported a number of tea plants. Using his horticultural expertise, he packaged them up in layers of soil, like a trifle, which he then encased in a glass box and arranged to be clandestinely transported to India. Within a generation new Indian tea plantations, centred around Darjeeling, were outstripping China both in quantity and quality.

It is only recently that a scientific explanation has begun to emerge as to why an infusion of the leaves of *Camellia sinensis*, one of about eighty species of *Camellia*, has been so successful in a world dominated by humankind. Like coffee (see page 283) tea contains the stimulant chemical caffeine. Therefore, all the stimulating 'pick-me-up' addictive effects associated with coffee are also present in tea. But tea has more to offer, which may help explain why it was originally cultivated with such enthusiasm so many thousands of years ago in the Far East, where it grew naturally in the wild. Tea's health-enhancing secrets are being revealed by research into theanine, a compound found only in tea plants and certain species of mushroom. This chemical crosses from the bloodstream into the neural pathways of the brain, where, according to medical trials conducted in 2007, it increases alpha brainwave activity, inducing a calmer yet more alert state of mind. When theanine is combined with caffeine, as it is in tea, the cognitive impact is much greater than with caffeine alone. Improved visual, numeric and verbal information processing and performance are just some of the observable benefits.

German scientists have recently discovered why Western tea drinkers may not benefit from the medicinal effects of drinking tea in the same way as Asian people, whose lifestyles and culture have evolved with tea over thousands of years in the Far East. When milk is added, many of tea's therapeutic properties are blocked. Compounds called caseins counteract tea's ability to increase blood flow around the body.

The success of tea, taken with or without milk, is an extraordinary story. At 3.5 million tonnes cultivated per year, the weight of the world's tea crop is about equivalent to cacao and half that of coffee (see page 283). Now add to that the ancient tabletop art of tasseography, the practice of divining tea leaves, and its place in the future looks even more assured . . .

Cannabis

> **FAMILY: CANNABACEAE**
> **SPECIES: CANNABIS SATIVA**
> **RANK: 54**

A psychoactive weed that captivated the imagination of
people long before the beginning of recorded history

WHAT COULD BE more innocent than young children enjoying the crazy sensation of spinning round in circles in the playground? A few years later they may seek the thrills of a high-speed roller-coaster ride that hurls them through the air on tracks twisted into the shape of a giant corkscrew. Bungee-jumping off a suspension bridge or parachuting out of an aeroplane at 3,000 metres appeals to others in adulthood. People of all ages, it seems, are naturally attracted – and some addicted – to the feeling of having their bodies driven beyond whatever margins may have been set by nature.

The desire to push human consciousness beyond its normal operating limits also helps explain how a particularly fast-growing weed became the most popular, and is probably the most ancient, recreational drug of all time. Tetrahydrocannabinol (or THC) is a psychoactive chemical produced by the plant *Cannabis sativa*, which has given human culture everything from pot to marijuana and hashish to bhang.

Angiosperms such as *C. sativa* (which belongs to the Rosid family) evolved in the Cretaceous Period (*c.* 145–65.5 million years ago) and have become some of nature's most brilliant protagonists in the development of chemical attack and defence. Evolutionary experts believe that THC was originally synthesized to fend off bacteria, insects and over-zealous herbivores. Its tactics are far more subtle than simply producing fatal toxins. Instead of poisoning its adversaries outright, cannabis plants disorientate, disable or bamboozle potential predators so they simply forget where the plant they nibbled came from in the first place. This has the added advantage of preventing a herbivore that happens to have natural immunity from cannabis toxins (a possibility afforded by sexual reproduction) from decimating plant populations.

When Stone Age humans experimented by eating the female flowers of *C. sativa* they experienced first hand the dazzling effects of THC on the human mind. Modern medical researchers were amazed to find that THC activates a vast number of receptors called CB_1 located in the human brain. In 1992 two Israeli scientists discovered that these receptors are used by a chemical called anandamide produced naturally in the brain. This chemical regulates short-term memory. Its job is to stimulate forgetfulness – especially important for liberating the mind from unpleasant recollections such as the pain of childbirth or severe physical trauma. Usually it achieves its objectives in a very short space of time, and is metabolized by the body in just a few seconds.

THC messes with this naturally occurring mental process. Not only does it invade neural pathways normally traversed by anandamide, but it lingers in the brain for several hours, as opposed to just a few seconds. The result can be literally mind-boggling, significantly exaggerating an individual's awareness of what is happening in the present, but temporarily blocking experiences from the past. This explains why smoking cannabis is renowned for making its users feel as if time has stopped and as if every sensation is extra-special. Music is said to peel apart with every voice and instrument momentarily isolated, standing out in solo. What are normally everyday objects become inherently more meaningful. Sensations such as taste and touch are similarly stopped in time

– as if digitized in slow motion – appreciated and enjoyed frame by frame in a zone of temporal suspension. Such sensual exhilaration causes anxieties about the future and worries about the past to disappear into blissful irrelevance . . . well, at least until the weed wears off.

This combination of increased psychological awareness of the present with the lack of toxic aftershock helps explain why cannabis extracts such as hashish and marijuana have been immensely popular drugs. THC in cannabis may have originally evolved as a protective mechanism to make herbivorous predators forget where the plant grew, but some time during the emergence of modern humanity its highly pleasant sensual powers almost inevitably resulted in deliberate cultivation for its extra-powerful sensory side-effects. Ancient shamans of the prehistoric hunter-gatherer world are known to have used cannabis in their herbal potions, concocted to induce a trance-like state for contacting the spirit world. Sacks of marijuana leaves discovered by archaeologists in China have been found buried next to mummified bodies dating back 3,000 years. From the earliest recorded religions such as Hinduism, to American churches founded in the twentieth century, cannabis has almost always played a ritualistic role in some form or other.

The worship of Shiva, most important of the Hindu gods, has been celebrated in northern India for more than 1,000 years by drinking bhang, a potent milky drink made from crushed cannabis flowers (see plate 7, bottom), also a favourite among Sikhs. Traces of hashish have been found in Nordic cultures, where it is thought to have been used in erotic festivals dedicated to their fertility goddess Freya. Even the Jewish Torah (Old Testament) is now widely thought to have condoned the use of *kaneh-bosm* (cannabis) as an essential component of the oil of anointment, prescribed by God as an initiation rite to be given to Hebrew priests and kings (see Exodus, chapter 30, verses 22–3). Despite the Muslim ban on intoxicating substances, the whirling dervish dances of

On Drugs

Sufi Muslims have traditionally been associated with the higher states of awareness brought on by the effects of cannabis. For the Rastafarian movement founded in Jamaica in the 1930s, the taking of cannabis is a fundamental ingredient in the spiritual experience of uniting the physical world with dimensions of divinity.

Secularism has not in any way suppressed the global appeal of cannabis. Despite its prohibition in the early twentieth century (along with most other recreational drugs, see page 314), smoking pot became a potent expression of peace, love and freedom following World War II. The association of cannabis with popular culture from modern jazz to the Beatles climaxed in the 1960s. Despite a 'war on drugs' waged by the US government since the early 1980s, cannabis is one of America's most lucrative cash crops.

Cannabis is grown today in as many as 172 countries, mostly indoors under bright lamps to avoid detection. An unintended consequence of modern prohibition has been the cultivation of cloned varieties that yield up to 25 per cent THC instead of the traditional 2–3 per cent from plants traditionally grown outdoors. The market is now flooded with new strains of highly potent marijuana such as Skunk, Northern Lights and California Orange, which some medical researchers believe could lead to far more significant long-term health issues for users, especially for those with pre-existing mental health problems. Governments today are therefore in a quandary. While the plant's universal appeal and ease of cultivation makes prohibition laws largely unenforceable and prohibitively expensive to put into effect, there are misgivings preventing the relaxation of the rules owing to new strains coming on to the market.

Cannabis succeeded in the ancient human world because its chemical properties helped people get closer to a spiritual nirvana that made sense of everyday life. It continues to succeed for some of the same reasons, providing many with a temporary escape from the anxieties of consumerism and modern urban living. Cannabis is a drug that stretches human consciousness beyond its normal limits, a more potent form of that feeling all of us first tasted as youngsters when spinning around in circles in the playground.

Tobacco

```
FAMILY: SOLANACEAE
SPECIES: NICOTIANA TABACUM
RANK: 67
```

A highly obnoxious plant that continues to decimate populations

ONE OF THE MOST celebrated moments in all human history was the accidental discovery of the Americas by a European mariner in 1492. Christopher Columbus collided with these unknown continents while attempting to find a new trade route to China. Columbus's colossal miscalculation of the actual distance to China had monumental consequences for global history. As many as twenty million indigenous Americans are estimated to have died as a result, mostly from smallpox. Today, a very small number of their descendants cling on to America's traditional way of life in small specially administered reservations.

Most accounts of American history tend to overlook the fact that the natives unwittingly inflicted their own much bigger holocaust on the invading Europeans. When sixteenth-century colonists began to settle in North America, earning their living by cultivating the land and trading overseas, little did they know that they were, quite literally, sowing the seeds for the premature demise of hundreds of millions of people all over the world. In the last 500 years the true human cost of the Native American people's gift of a plant belonging to the same family as the common potato (see page 210) has come to haunt the modern world. The worldwide adoption of growing and smoking tobacco – a Native American habit – has since led to the death of billions of people.

Nicotiana tabacum is one of the most influential plants in human history because of its highly addictive power and devastating effects on human health. Nicotine, its most important active ingredient, is a naturally occurring insecticide that is especially concentrated in the sixty-four species of plants in the genus *Nicotiana*. Once inhaled, the compound takes just ten seconds to travel through the bloodstream and cross into the brain. Here it binds to various receptors making people feel relaxed, contented – even euphoric – while also stimulating the release of glucose, boosting energy levels, raising the body's metabolism, enhancing concentration and increasing overall levels of alertness. However, it is the *absence* of these effects that causes people who smoke to feel a desperate urge to have another dose. Nicotine is metabolized by the body within two hours and its addictive properties are driven by the urge to restore its mental and physiological effects.

Studies on addiction suggest that a common feature of human consciousness is the habit of dividing up time into manageable segments. Lighting a pipe or smoking a cigarette satisfies this natural inclination, inserting a series of pleasurable commas into everyday life. A smoking addict finds that, in the absence of a nicotine fix, time drags on *ad infinitum*, sinking deprived individuals into a depressive abyss. The brain's response to a flood of nicotine is to build additional chemical receptors, so developing tolerance to the drug. But when the drug is absent, the sensitivity of all these neurons causes severe withdrawal symptoms. These extra neurons are more or less permanent.

But, addictive as it is, nicotine does not kill. Large doses administered through the skin could be lethal, but unlike users of cocaine (see coca, page 312) or LSD (see ergot, page 317), no cigarette smoker has ever died from an overdose of nicotine. Levels in tobacco are simply not high enough. Diseases such as cancer, emphysema, or cardiovascular ailments such as heart attack and stroke, are statistically much more likely to occur in smokers but they have nothing to do with nicotine. As many as nineteen different cancer-provoking chemicals have been isolated in cigarettes, the most lethal of which are radioactive isotopes of lead and polonium that rip through living cells, shredding them from the inside out.

Native Americans were regarded by most Europeans as savages whose habits and culture included everything from cannibalism to devil worship. Despite these prejudices, within 150 years of Columbus's arrival tobacco plants and the Native American practice of smoking through a pipe had spread via European explorers throughout the world. Other cultures, it seems, were just as susceptible to the charms of tobacco. Spanish explorers transplanted the crop to the Philippines in 1575, from which it quickly spread to China. At the same time Portuguese traders took tobacco to Africa, India, Java and Japan. By 1621 smoking tobacco was widespread in the streets and mosques of Persia, imported via traders in Baghdad. By 1700 the Persians had their own tobacco-growing industry, and were exporting as many as four thousand water pipes to smokers in India each year. By the end of the seventeenth century tobacco was, by any definition, a global crop, which had spread, says one historian, '*like ripples from a handful of gravel tossed into a pond*'.

Norfolk farmer John Rolfe was one of the plant's chief protagonists. He took the seeds of sweet-tasting *N. tabacum* from the Caribbean and planted them along the banks of the James river in Virginia in 1612. Virginian tobacco quickly became the world's favourite blend. At the beginning of the seventeenth century annual UK imports of North American tobacco totalled just 114,000 kilograms. One hundred years later they had risen to seventeen million kilograms.

Tobacco's biggest break came about because of a contest. Until 1870 it took a skilled manual worker one minute on average to hand-roll four cigarettes. By this time the worldwide demand for American cigarettes was so great that Virginian cigarette-making company Allen & Ginter offered an enormous cash prize ($75,000) to anyone who could automate the process. James Bonsack, a humble

American inventor, seldom makes it into history books, but surely his achievement was as monumental as that of any king or emperor. When Bonsack's automatic cigarette-rolling machine cranked into action in September 1880 it could process a staggering 12,000 cigarettes an hour, as compared to the yield of a manual worker, who averaged 240 in the same time. By 1895 North America's cigarette output had increased from sixteen million to *forty-two billion* a year. Two world wars, mass-production and a series of carefully targeted marketing campaigns that glamorized smoking for women all contributed to tobacco's exponential rise in popularity in the first half of the twentieth century (see plate 12, bottom).

Then, one day in 1950, a British medical expert threw a monkey wrench into the works. It took the British government seven years to digest Richard Doll's conclusive report that linked smoking tobacco with lung cancer. From 1957 onwards, government advice began to inform the public that smoking, however enjoyable, was potentially disastrous for human health.

In tobacco's American birthplace, coming to terms with the knowledge that smoking causes ill health (reducing life expectancy on average by at least ten years) was, like coming to terms with the ban on slavery, extremely hard for many people. Corporations addicted to commercial profit (not nicotine) fought battles with consumers and campaign groups to disprove the link, or at least make sure any proof was inconclusive. A conspiracy called Operation Berkshire involved the collusion of America's biggest tobacco companies in an effort to forestall any federal legislation to limit their business.

However, concerted government intervention and public health awareness programmes have now resulted in a decline in tobacco smoking in many developed countries. Today only 21 per cent of the US adult population are smokers compared with 42 per cent in 1965. However, in the less-developed world (especially Africa and Asia), where governments are less able to raise awareness among their people or more open to exploitation by giant tobacco corporations, the total number of smokers still increases daily. According to a report by the World Health Organization published in 2002, tobacco consumption in developing countries is rising at an average rate of 3.4 per cent per year.

The same report also reveals that the impact of *N. tabacum* today, thanks to its highly addictive insecticide, is every bit as pronounced, despite most people knowing the consequences to their health. It includes the following chilling statistics: ten million cigarettes per minute are sold across the world every day; twelve times more British people have died from smoking than were killed in World War II; $150 billion a year is the cost of smoking-related healthcare in the United States alone.

Yeast

> FAMILY: SACCHAROMYCETACEAE
> SPECIES: SACCHAROMYCES CEREVISIAE
> RANK: 8

A microscopic fungus with a macro-economic impact throughout all human history

A CURIOUS CORRELATION seems to exist between the size of a living thing and its potential effect upon planet, life and people. Generally, it seems, the smaller a life-form the greater its impact. The Oomycete water mould (see page 40), cyanobacteria (see page 20) or smallpox (see page 13) are all good examples – as is the extraordinary microscopic fungus commonly called yeast. There are roughly *twenty billion* individual cells in just one gram of yeast. Each cell is capable of performing the equivalent of biblical miracles – for example turning sugary fruit juice into wine – a chemical metamorphosis that has dramatically influenced the culture of human civilizations.

More than 1,500 yeast species have been identified by scientists to date, although by some people's reckoning these represent only about 1 per cent of those that exist in nature. These microbes feed on sugars and thrive in every environment – land and sea – spreading when necessary as spores in the air. But there is one species in particular, *Saccharomyces cerevisiae*, that has played a profound role in history. It is almost exclusively thanks to the action of this single-celled microscopic fungus that humanity has been able to enjoy everything from leavened bread to fine wine. What's more, some of our best prospects for fuelling sustainable industrialization and transportation systems in the future are derived from by-products associated with this quasi-miraculous yeast.

Yeasts make their living by digesting sugars (such as glucose, fructose or sucrose) to make energy. The waste products of this process, which is called fermentation, are carbon dioxide gas and a liquid compound called ethanol. It is these waste products that have been of such enormous value to humans.

Whoever was the first person to discover how to bake bread will forever remain a mystery, but painted hieroglyphs show that baking was widely practised by people living in the civilization of ancient Egypt (see also pages 203). Someone probably left out a mixture of flour and water on a warm day and this then became contaminated by *S. cerevisiae*, a naturally occurring yeast that frequently grows on fruit skins. This doughy mixture would then have fermented as the yeast began to break down sugars in the flour, causing it to rise with the output of carbon dioxide gas. Heat from a baker's oven encapsulates this sponge-like texture, turning soft dough into the light fluffy bread we are so familiar with in our everyday lives. Meanwhile, the ethanol would have evaporated in the heat of the oven, but not without leaving behind the irresistible aroma of a freshly baked loaf.

Such smells have had ubiquitous appeal throughout human history, and the art of baking leavened bread, begun in ancient Egypt, became that civilization's most precious export. Everyone around the Mediterranean soon wanted grain (cultivated alongside the river Nile) with which to bake their own bread. Each time, it seemed, a miracle occurred in the oven as uncooked but fermented dough metamorphosed into a wonderful crispy loaf. Certainly it seems that the Israelites enjoyed leavened bread during their captivity in Egypt since the Bible tells us that God commanded them during their exodus to celebrate their release by the taking of unleavened bread. It was meant as a permanent reminder that they were in such a rush to leave that there was no time to wait for the dough to rise.

Yeast's other miracle has been no less dramatic. Ethanol (a type of alcohol) is a colourless liquid used to make different types of intoxicating drinks – the most famous of which are beer, wine and spirits (such as gin, whisky or rum). When mixed with acidic fruits, *S. cerevisiae* ferments sugars in a way that seems to have an especially pleasing effect on the human palate, producing fragrant chemicals called esters. Other yeasts such as *Zygosaccharomyces* or *Brettanomyces* produce off-putting antiseptic flavours. These organisms sometimes illicitly creep into the fermentation process, particularly through the use of old barrels, much to the dismay of wine makers throughout the ages. *S. cerevisiae* (often called brewer's or baker's yeast) has no such nasty side-effects, which is why strains from this species of yeast have most commonly been used for producing almost all the alcoholic drinks ever consumed by man.

Like baking leavened bread, no one quite knows when it was that people sipped their first cup of wine or draught of beer. Some of the earliest evidence of wine production goes back 6,500 years to Georgia, in central Asia (see also grape, page 307). It then seems to have spread throughout Mesopotamia and northern Greece. Ancient Sumerians (*c.* 5000 BC) are known to have set aside as much as 40 per cent of their annual grain harvest for the production of beer (hops began to be used only from medieval times). Before that no one really knows if hunter-gatherer humans ever enjoyed the dizzying by-products of *S. cerevisiae*.

What is certain is that drinking alcohol has caught on throughout human history, even more than smoking nicotine – and has proved just as addictive. Ethanol, like THC (see page 292), is a psychoactive drug. In small quantities it produces feelings of relaxation and cheerfulness, usually provoking behavioural changes such as loss of inhibitions and an increase in self-confidence. As blood-alcohol levels rise, it depresses the central nervous system leading to feelings of lethargy. High doses cause confusion, dizziness, stupor, coma and sometimes death. Most often too much alcoholic drinking ends in

On Drugs

unconsciousness, which has the advantage, unlike some other drugs (see coca, page 312), of breaking its immediate administrative cycle.

Alcoholism today does not account for as many deaths as nicotine, but its impact on society is no less devastating. The World Health Organization's *Global Status Report on Alcohol (2004)* provides a sobering snapshot. As many as sixty different diseases and injuries are caused by alcohol consumption of which the most pernicious are cancers, suicide and motor accidents. Nearly two million deaths a year are alcohol-related (3.2 per cent of all deaths) – a third of which are caused by the drug's stupefying effects – called 'unintentional injury'. Global patterns of consumption are uneven, however.

In some Muslim countries where the sale of alcohol is banned consumption is extremely low. More than 95 per cent of the adult population in Iran, Saudi Arabia and Egypt are teetotal. In other countries, however, despite attempts to outlaw alcohol (such as US Prohibition 1919–33) the picture is reversed. Teetotallers are especially rare in places like the UK (12 per cent), Norway (6 per cent) and Germany (5.1 per cent). Such patterns help explain why as many as 55,000 Europeans aged between nineteen and twenty-nine die each year as a direct consequence of alcohol abuse.

The by-products of the fermentation process of *S. cerevisiae* are not all alcohol-related, however. A yeast extraction process pioneered in the nineteenth century by German chemist Justus von Liebig (1803–73) makes a nutritious vitamin-rich edible paste from the left-over sediment of brewer's yeast. Marmite (love it or hate it) is especially popular in Britain, which has been manufacturing this vegan health food since the early 1900s. Other potentially positive uses for the by-products of ubiquitous *S. cerevisiae* are represented by the recent interest in biofuel as a potentially sustainable means of fuelling the modern economy. Ethanol, which can be used to power everything from cars to power stations, is usually made by fermenting crops of sugar or maize. As many as 6.8 million cars in Brazil are now able to run on ethanol fuel, which is available in 35,000 of the country's petrol stations. To what extent such practices can be made genuinely sustainable without destroying the world's rainforests, increasing food prices or using so much fuel and fertilizer in the farming and harvesting of crops that the environmental benefits of biofuel become purely political/academic achievements, remains to be seen.

Brazil is now being rapidly followed by other nations such as the USA. Fuel from ethanol looks set to ensure that the highly significant story of how a microscopic, single-celled fungus transformed the eating and drinking habits of human civilizations continues to power on into the twenty-first century and beyond.

Grape

> **FAMILY: VITACEAE**
> **SPECIES: VITIS VINIFERA**
> **RANK: 73**

A species that has benefited from both fungus and man,
turning its innocent juice into a drink fit for the gods

VITACEAE IS A family of about sixty species of flowering trees. Among them are several that bear grapes – juicy, sweet fruits that are also home to naturally occurring yeasts (see page 302). But it is the grape from a single species – *Vitis vinifera* – that has had such a powerful impact on world history and human culture.

More than 70 per cent of all grapes grown today are pressed and fermented into wine – an alcoholic drink enjoyed particularly by the inhabitants of Europe, the Americas, Australia, New Zealand and South Africa. Plantations of *V. vinifera* now inhabit every continent in the world except Antarctica. Latest figures estimate that the worldwide area dedicated to cultivating grapes extends to about 76,000 square kilometres – nearly twice the size of Switzerland – and the rate of cultivation is increasing by 2 per cent per year.

The human love affair with grapes has a great deal to do with that naturally occurring yeast, *S. cerevisiae* (see page 302), which coats the skins of the fruit with what looks like a dusting of light white powder. It can only be supposed, rather like the discovery of how to bake bread, that someone left grape juice out in the sun and the process of fermentation began. Before the liquid turned into sour-tasting vinegar, they must have drunk it and discovered the pleasing effects of ethanol on body, mind and spirit. Vineyards need full-time attention; processing grapes takes time and keeping wine from going sour requires technology in the shape of air-tight containers, something possible only with the invention of impermeable pottery jars capped with sealing wax. Innovations such as these emerged in central Asia and the Middle East alongside the first settled societies beginning approximately 8,000 years ago.

The success of the grapevine *V. vinifera* is therefore synchronous with the rise of human civilization. A freak of nature may have given this vine a helping hand. About 5,600 years ago waters from the Mediterranean burst through the Bosporus, flooding thousands of square miles of land, raising the region that is now the Black Sea to present day sea levels. This was where it is thought the first vine-growing people may have lived, in the proximity of where *V. vinifera* grows in the wild. Survivors of the flooding of the Black Sea pushed their way to the east, south and west, taking with them expertise in how to grow vines and turn grape juice into wine. Such a sequence of events also helps explain why the word 'wine' is so strikingly similar across such a wide range of distantly related languages from the Hittite *wiyana* to the ancient Greek *oinos*, Arabic *wayn*, Georgian *gvino* and Latin *vinum*.

The ancient world's premier vintners – and those most in love with the taste, culture and after-effects of drinking wine – were undoubtedly Greek. These people lined their wine jugs with tree resin as a natural preservative, antibiotic and insecticide (see page 95). Greek wine called *retsina* still retains this ancient flavour. Wine was the drink of the gods. The Greeks even had their own wine god, Dionysus.

Although the concept of processing grapes into wine for the pursuit of politics and pleasure was first practised by the well-to-do of ancient Greece, it was the Romans who spread winemaking to all parts of their imperial domains and seeded the first vineyards in Spain, France and Germany. The Romans had their own copycat version of the Greek symposium, which they renamed *convivium*

– from which derives our modern word 'convivial' – used to describe a social occasion loosened by wine. They worshipped Dionysus, the god of wine, but renamed him Bacchus.

Cultural fusion between divinity, fine wine and high society was reflected in the emerging religions of the Mediterranean world. Pagan worship of wine through the gods is not culturally so very different from the idea of wine turning into the Blood of Christ during the Catholic sacrament. The Holy Eucharist, still celebrated today by hundreds of millions of Christians across the globe, requires the drinking of wine to commemorate the death on the cross of the son of God, Jesus Christ, who is recorded by St Paul in his letter to the Greek Corinthians as saying: '*The cup of blessing which we bless, is it not the communion of the Blood of Christ?*'

Following the decline of the Roman empire the market for wine collapsed although the link between wine and divinity lived on. From *c.* AD 500 to 1000 Europe's premier vintners were mostly monastic. Benedictine monks had to maintain healthy vines in order to supply the wine needed to celebrate their holy sacrament. Vineyards were also a handy way of earning extra cash from trade with European aristocracies, which gradually began to reassert themselves from the ninth century AD under the auspices of the feudal system. By the fourteenth century the thirst for wine among the royalty and aristocracy of England was insatiable. From 1305 to 1309 nearly 300,000 barrels of wine were exported by sea from Gascony, in France, to England – that's *900 million litres* a year. When Europe's conquest of the Americas turned to colonization from the sixteenth century onwards, settlers took their vines with them. Jesuits from Spain established their own vineyards to secure sufficient local supplies of eucharistic blood.

Artificial selection constantly runs the risk of rubbing up against nature's systems. A lack of genetic variety in *V. vinifera*, like the potatoes brought back by colonists from the New World (see page 210), meant that vineyards full of genetically identical grafted fruits were extremely vulnerable to attack. And when the predator finally arrived, the effects were devastating.

Monsieur Borty was a French wine merchant and is credited by history with importing a number of American vines, which he planted somewhere along the banks of the river Rhône in the summer of 1862. Little did he know that their roots were infested with tiny yellow venomous aphids that had survived the Atlantic crossing, probably because new-fangled transportation by steamship had dramatically reduced the time it took to travel from America to Europe. So prolific is the reproductive capacity of this aphid species, *Daktulosphaira vitifoliae*, that a single female can produce as many as 25.6 billion descendants in just eight months – all without the assistance of a male. These aphids suck out sap from the roots of *V. vinifera* and replace it with their own specific cocktail of venom. No one knew the cause of the mysterious vine-killing disease called *phylloxera* until one day in 1868, when French biologist Jules-Emile Planchon came up with his aphid-infestation theory after some ramblers told him they had spotted the insects nonchalantly marching along some vine roots *'strolling along like good bourgeoisie going into a restaurant with walking sticks in their hands'*.

So desperate was the French government that it offered a 320,000-franc prize to anyone who could find a way to kill the pest. While organic chemists tried every conceivable type of pesticide, French winemakers resorted to experimenting with everything from cow's urine to tobacco and elder leaves to crushed bones dissolved in sulphuric acid. Nothing seemed to kill the happy aphids and within fifteen years the French wine industry had collapsed to half its former size. By 1875 more than 40 per cent of all French vineyards had been destroyed – some 2.5 million acres – with another 1.5 million infested.

An American species finally saved the French wine industry from annihilation – and with it revived the fortunes of *V. vinifera*. Two French wine growers, Leo Laliman and Gaston Bazille, suggested grafting the European winemakers' *V. vinifera* on to the roots of American *Vitis* species immune to the aphid plague, thus restoring some sense of genetic diversity. Despite entrenched opposition from French vintners, who were paranoid that such dramatic surgery would impair the delicate flavours of their high-class wines, trials of this hybridization process proved successful.

From the mid-1870s on, the enormous task of uprooting, grafting and replanting every vine in France got under way, a process that came to be known as the 'reconstitution'. Today there is hardly a vineyard of *V. vinifera* anywhere in the world (except Chile) whose rootstock is not derived from a different Native American species. Since roots simply channel nutrients and water up from the ground, no difference in flavour has ever been detected in this now ubiquitous hybrid species since the chemistry of turning water into grape juice is still undertaken by the genes of *V. vinifera*, which are able to thrive, aphid-free, safely above the ground.

There is as yet no known way of eradicating these aphids, so the French prize remains locked in a bank vault, still waiting to be claimed. It was judged that the solution pioneered by Monsieurs Laliman and Bazille did not, strictly speaking, lead to the eradication of the pest. Despite its brush with catastrophe in the hands of nineteenth-century cultivators, *V. vinifera* is now a global success story.

Coca

> FAMILY: ERYTHROXYLACEAE
> SPECIES: ERYTHROXYLUM COCA
> RANK: 96

Highly addictive leaves that flavour the world's most popular drink

THE STEPAN CORPORATION, an American chemical company located in New Jersey, imports more than one hundred tonnes of coca leaves a year. Usually trade in coca is illicit thanks to the highly addictive compound that is often extracted to make a recreational drug called cocaine. But Stepan's importation is for an entirely different use. It processes the leaves in vast vats on its nineteen-acre industrial estate in order to extract a cocaine-free non-narcotic flavouring that can be added to a recipe for the world's 'favourite' soft drink, named after the plant itself: Coca-Cola.

For at least 2,500 years before European explorers first discovered the Americas in the late fifteenth century, South American people had enjoyed chewing coca leaves. The Incas (1438–1533) considered the plant a gift from the gods thanks to its stimulating, energy-enhancing effects. In traditional native South American cultures a wad of coca leaves is tucked between the gums and the cheek, where it is gently sucked and squeezed in the mouth for up to three hours at a time. This ancient technique allows a low but constant dose of cocaine, an alkaloid drug, to filter into the bloodstream where – like caffeine – it acts as a stimulant, restricting blood vessels, increasing the heart rate and readying the body for action. Consuming cocaine in this measured way avoids any sudden rush of the drug to the brain. Modern-day effects, such as mind-blowing euphoria, addiction, withdrawal and occasionally death, are largely due to the consumption of cocaine in a purified form via injection or inhalation.

When Spanish conquerors first came across coca-chewing they banned the practice, dismissing it as primitive and un-Christian. But they changed their minds when the locals refused to work in the silver mines – such as those at Potosí, which by the 1650s had begun to make many Europeans extremely rich. Payment in coca leaves provided sufficient incentive, it was found, for slaves to extract all that silver and gold on behalf of their European oppressors.

For more than 200 years people outside South America remained blissfully ignorant of the secrets of what coca could do. Then, in 1858, an Italian doctor called Paolo Mantegazza (who regularly corresponded with Charles Darwin) returned from practising medicine in South America to take up the post of surgeon-general at a leading hospital in Milan. His paper entitled 'On the Hygienic and Medicinal Properties of Coca', published the following year, exposed the euphoric effects of chewing coca leaves, which he had discovered for himself after spending time among the indigenous natives of South America: *God is unjust because he made man incapable of sustaining the effect of coca all life long. I would rather have a life span of ten years with coca than one of 10,000,000,000,000,000,000,000 centuries without . . .*'

His words triggered a race among European medics to find the secrets of these wonder-leaves. Within a year, Albert Niemann, a graduate student at Gottingen University in central Germany, had written a dissertation explaining how to isolate the compound cocaine from the leaves of coca. Three years later a Corsican pharmacist called Angelo Mariani patented a new drink, made by mixing wine from Bordeaux with cocaine from coca leaves. Even the Pontiff himself, Pope Leo XIII (1810–1903), took fondly to the convivial effects of cocawine, endorsing the drink by awarding a gold medal to Angelo Mariani for *services to the benefit of humanity* (see plate 10, top). Over the next twenty years his cocktail became an international hit, eventually attracting the attention of morphine addict John Pemberton, the man who in 1885 developed his own American cocawine alternative, which he later called Coca-Cola.

Swig back a mouthful of Pemberton's original French Wine Coca today and you'd get quite a kick; the original recipe contained up to six milligrams of cocaine per fluid ounce of wine. Pemberton also added caffeine from the kola nut as an additional stimulant, the combined effects of which would most certainly have lived up to his claims that his coca drink would benefit *scientists, scholars, poets, divines, lawyers, physicians, and others devoted to extreme mental exertion* as well as being *a most wonderful invigorator of the sexual organs*.

The nineteenth-century love affair with the coca plant continued with cocaine tooth drops, marketed in the US as a family-friendly solution to toothache (see plate 11, bottom). Meanwhile a new technique for removing cataracts was pioneered by German surgeon Carl Koller in 1884, who showed how an application of cocaine could numb the cornea, providing blessed relief in an age when cataract removal was likened to having an eye punched through with a red-hot needle. Until 1916, syringe kits, like those used by fictional super-sleuth Sherlock Holmes in Arthur Conan Doyle's detective stories, were sold as lifestyle accessories in upper-crust shops such as Harrods of London.

But by the early twentieth century Western governments, paranoid about the idea of their troops getting high on drugs, began to wake up to the epidemic of cocaine addiction that was sweeping through American and European societies (see also heroin, page 323). The International Opium Convention of 1912 led to a general prohibition on drugs, with countries such as the US following up with their own legislation, such as the Harrison Act of 1914, which banned the sale of heroin or cocaine except under licence.

The change of mood had begun as early as 1903, when the secret recipe for Coca-Cola was surreptitiously downgraded to remove coca's psychoactive ingredients. From this time on only spent coca leaves, with most of their active compounds removed, were used to flavour the drink, although its caffeine content and some molecular traces of cocaine remain. Despite the removal of its main psychoactive ingredients, this energy-giving sweet-tasting drink had worldwide appeal, furthered by some exceptionally smart marketing tactics by the company's chief Robert Woodruff (1889–1985), who, among other things, created the image of Santa Claus still familiar in many parts of the world today in his red and white costume.

Efforts by Western governments to eradicate cocaine as a recreational drug were undone during the 1970s when the plant underwent a dramatic renaissance to become one of the world's most widely used addictive substances. Today more than 150,000 hectares of coca are cultivated deep in the forests of Colombia, Peru and Bolivia, supplying an estimated fourteen million addicts world-wide. Most of it is grown illegally and turned into cocaine powder in small makeshift factories that are re-established as fast as the authorities locate and destroy them, although 30,000 hectares of coca are legally cultivated in Peru for medicinal purposes. The government is trying to revive coca tea as a safe, traditional way of harnessing the plant's natural medicinal qualities and as an antidote to altitude sickness for visiting tourists.

Cocaine's modern revival began among America's upper-middle classes. In 1974 the *New York Times* magazine called it 'the champagne of drugs'. Its popularity surged a decade later after someone discovered that by mixing cocaine powder with sodium bicarbonate the drug could be inhaled in a smokable form giving a more instant, addictive high. The spread of this variant, called 'crack' after the sound it made during manufacture, was boosted by a combination of severe economic dislocation in South America, giving birth to cartels of growers and a large social underclass in America's sprawling inner cities, which became the drug's urban sales force. In its new form, the drug spread to Europe, where its largest communities reside in Spain, Italy and the UK. New markets for the drug are growing in other European countries and west Africa.

There is an intriguing postscript to the story of the coca plant's success, both through its use as a highly addictive recreational drug and as a flavouring for 'pick-me-up' fizzy drinks. See how drugs derived from and grown in the western hemisphere are mostly *stimulants* (such as nicotine and cocaine) while those that dominate in the eastern hemisphere are *depressants* (opiates such as heroin and morphine, see poppy, page 321). Note how well matched are the characteristics of these contrasting drugs with the traditional cultures of East and West. The developed Western world's thrusting, violent consumerism and urge for physical and commercial conquest contrasts with Far Eastern cultures' traditional desire for self-sufficiency (Japan until 1868, China until 1975 and North Korea still today) and spiritualism (such as the spread and appeal of Buddhism throughout Asia).

When Japan sprang out of the blocks in the late nineteenth century in a bid to bring Western industrialization to the Far East, its then colony Taiwan was sown with coca plantations – a counter-cultural drug – to feed the mother country. To what extent did Japan's shift to recreational stimulants (aided by Dutch industrialists) reflect or cause its cultural and economic transformation? To what extent are human cultures the outward expressions of powerful plants that just happen to grow in their gardens?

Ergot

FAMILY: CLAVICIPITACEAE
SPECIES: CLAVICEPS PURPUREA
RANK: 69

A parasitic fungus that has helped colour human culture from ancient Greece to The Who

ALBERT HOFMANN WAS extremely fortunate not to have fallen off his bicycle one April evening in 1943. In a journal, the Swiss chemist from Basel described the most terrifying ride of his life:

> *'On the way home . . . everything in my field of vision wavered and was distorted as if seen in a curved mirror. I also had the sensation of being unable to move from the spot. Nevertheless, my assistant later told me that we had travelled very rapidly. Finally, we arrived at home safe and sound, and I was just barely capable of asking my companion to summon our family doctor.'*

Hofmann, who died in 2008 at the grand age of 102, was describing his experience of ingesting a drug he had synthesized in a search for a new type of medicine. He called it lysergic acid diethylamide, more commonly known today as LSD. Hofmann later wrote how his bike ride was followed by at least six hours of extraordinary experiences in which even the tables and chairs in his room began to threaten him: *'Everything in the room spun around, and the familiar objects and pieces of furniture . . . were grotesque, in continuous motion, animated, as if driven by an inner restlessness.'* The ingredient in LSD responsible for

317

these bizarre experiences comes from a particular species of parasitic fungus known about since ancient times. It is called ergot.

Claviceps purpurea (the scientific name for the most common species) feasts on fields of rye, barley and wheat. When one of its spores lands on a floret of grass it mimics the behaviour of pollen by growing a tube that burrows its way down the flower's style. At the bottom it invades the flower's ovary and, virus-like, hijacks its seed factory to produce its own hard-cased reproductive apparatus called a sclerotum. This grain-like growth eventually falls to the ground, waiting for just the right conditions to fruit into a series of microscopic mushrooms that generate new spores that can be blown in the wind to begin its life-cycle all over again.

Tucked inside the ergot's sclerotic sack is a series of chemicals manufactured by the fungus to deter grazing animals from interfering with its grass-subverting routine. These substances, called ergotamines, make animals (including humans) temporarily mad. They were also the base ingredients used by Hofmann to manufacture LSD. The human and animal after-effects of eating ergot-infected rye, barley or wheat range from uninhibited convulsions to life-threatening gangrene. All domestic animals are susceptible, especially cattle. Since the discovery of LSD, a number of scholars and medics, including Hofmann, have begun to re-interpret key historical events with the hindsight of a better understanding of ergotamine drugs. For example, ergot is thought to have supercharged the climax of ancient Greece's famous Eleusinian mysteries, a late-summer festival celebrated in a temple dedicated to fertility goddess Demeter at the shrine of Eleusis near Athens. During this annual ritual as many as 3,000 pilgrims drank a secret drink called *kykeon* concocted from a range of herbs, fermented wine and barley. Experts believe ergot from infected fields nearby was probably responsible for the drink's notorious LSD-like hallucinogenic effects.

Indeed, the effect of ergot on human history – especially in cults concerning witchcraft and satanism – may have been a great deal more significant than surviving records suggest. In medieval times a fury called *ignis scaer* (holy fire) was known to cause wild behaviour among people in England, France and Germany. Several episodes have been recorded between the fourteenth and eighteenth centuries in Europe in which thousands of people danced and screamed wildly for no apparent reason. On one such occasion, in Aachen, Germany on 23 June 1374, it is recorded that people came out on to the streets and twisted and contorted their bodies in a wild uncontrollable manner. In July 1518 what has been described as a 'dancing plague' broke out in Strasbourg, France, where as many as 400 dancers took to the streets for up to a month, some of whom died from exhaustion or heart attack. Contemporary physicians ruled out supernatural causes, concluding it to be some kind of 'natural disease'.

The notorious hanging of fourteen women and five men in the early Puritan settlement of Massachusetts in May 1693 has also been attributed to the effects of ergot ingestion. Following the Salem witch trials these people were sentenced to death for behaving 'beyond the power of epileptic fits or natural disease', including screams, contortions and throwing objects. Descriptions of their behaviour, suggested by the local village doctor to be caused by supernatural forces, have since been likened to ergot poisoning.

Outbreaks have become much rarer since medics first began to understand the effects of this fungus on the human mind and body in the mid-1850s. Farmers became increasingly aware of the danger of parasitic infection by ergot towards the end of the nineteenth century and started to manage their crops using fungicides based on copper sulphate solutions, crop rotation (ergot survives for only about one year in the soil) and grain cleaning techniques. However, owing to Hofmann's discovery of how to synthesize LSD, ergot made a dramatic and unexpected comeback in the middle of the twentieth century. Since then its impact on the development of modern Western popular culture has been profound.

Right up until his death in April 2008, Hofmann believed LSD could have powerful remedial effects for people with mental health problems – as a potential cure for alcoholism and even to help inner-city people who have lost all contact with nature to reconnect with the universe. Indeed, LSD was a perfectly legal drug until the late 1960s. During this period extensive medical trials in both the US and UK were conducted by physicians, research students and even the American intelligence forces.

Meanwhile, LSD trail-blazed new forms of cultural expression in the 1960s based on the liberating effects of psychedelic 'trips'. Literary works such as Ken Kesey's *One Flew Over the Cuckoo's Nest* (published in 1962), about the fate of patients in an American mental institution, were written under the influence of LSD. Popular music by bands from the Beatles ('Lucy in the Sky with Diamonds') to the Moody Blues symbolized the rejection of the Establishment and its views on nuclear war, homosexuality and race. LSD, made widely available by the psychiatric profession, was a trigger that seemed to unleash a new confidence in ordinary people, giving them the courage (or the gall) to raise two fingers to those in positions of authority.

In 1970 LSD was classified as a Schedule One drug by the US government, with Timothy Leary branded by Richard Nixon '*the most dangerous man in America*'. A year later the United Nations Convention on Psychotropic Substances was agreed, banning LSD and almost every other conceivable hallucinogenic substance for legal use. Meawhile, recreational use went underground and an age of popular resistance to the political interference of the state in people's private lives, which continues to this day, was ignited by phenomena such as psychedelic rock performed by bands ranging from Pink Floyd to The Who.

Poppy

FAMILY: PAPAVERACEA
SPECIES: PAPAVER SOMNIFERUM
RANK: 79

Nature's most addictive narcotic has administered and relieved pain and pleasure in roughly equal measures

OVER THE LAST 300 years poppies have not only determined the fate of political and economic relations between the world's western and eastern hemispheres, they have also relieved countless humans worldwide from one of the greatest sources of suffering – excruciating pain.

The therapeutic effects of opium, dried extracts from the seed pods of the poppy species *Papaver somniferum*, have been known to people living in and around the Mediterranean for at least 6,000 years. Poppy heads have been found in a cave in Spain dating back to 4200 BC, which, according to experts, were probably used by Neolithic people for their intoxicating effects. The poppy's pain-relieving properties were well noted by the Greek medic Hippocrates (460–370 BC), the Roman physician Galen (AD 129–200) and the Persian polymath and scholar Avicenna (AD 980–1037). They all prescribed the white latex contained inside poppy seed pods to relieve pain and suffering.

Codeine and morphine, two active compounds found in poppy latex, are brilliantly effective at blocking pain receptors in the brain. Even today they are among the most commonly administered medicines in the world, essential for everything from relieving pain from toothache to providing comfort and relief to those suffering terminal cancer. But while the enormous medical impact of poppies is beyond dispute, the political and economic ramifications of its recreational use as a highly addictive pleasure-enhancing narcotic are a lot more ambiguous. The legacy of opium and its semi-synthesized derivative heroin is the mistrust and resentment that still simmers menacingly beneath the surface of relationships between East and West.

Until the nineteenth century the relationship between opium poppies and human beings was mostly positive on both sides. The drug was uniquely suited to helping settled people deal with their increasingly civilized, urban ways of life. Medics found that all manner of conditions from anxiety, boredom, fatigue and insomnia to squalling babies and diarrhoea could be better managed and controlled with the use of opium. As a result Arab conquerors in the eighth century took poppy seeds with them as they travelled throughout Asia, spreading the therapeutic, painkilling properties of the poppy's dried extracts to India and the Far East. Islamic cultures preferred remedial opium as an alternative to alcohol, which was proscribed by the Prophet Mohammed.

A Swiss-German alchemist called Paracelsus (1493–1541) was the first European to discover that the dried extracts from poppy latex (opium) could be dissolved in alcohol. He called his mixture laudanum (from the Latin *laudeo* meaning 'I praise') because of the satisfying effects it seemed to have on the mind. Paracelsus's potion was picked up by English doctor Thomas Sydenham (1624–89), who described its effects as '*a most delicious and extraordinary refreshment*' which he compared with '*that pleasure which modesty prevents the name of*'. This cocktail, known as an opium tincture, was then used increasingly by everyday people for everything from pleasure enhancement to pain relief and as a sleeping draught. Even into the late nineteenth century a wide range of prescription-free exotic opium-based potions were being promoted in UK pharmacists, such as 'Mrs Winslow's Soothing Syrup', heralded as the perfect solution for soothing a baby's toothache (see plate 10, bottom).

On Drugs

Two technological developments upset the use of opium as a popular oral medicament. The first was the invention of the hypodermic syringe by Scottish physician Alexander Wood in 1849. Directly injected drugs bypass the body's digestive processes, which normally metabolize harmful chemicals into less damaging derivatives before they reach the bloodstream and the brain. The effect of a drug shot straight into the vein is therefore substantially more powerful than when administered orally, crossing the all-important blood–brain barrier within seconds and releasing a highly potent 'rush'. The human body evolved over millions of years to deal effectively with ingested toxins (sickness and diarrhoea are two evolutionary defences). It has not yet had time to evolve a way that deals with a surprise injection.

The second technical change was the accidental synthesis of an acetylated form of morphine that turned out to be at least twice as potent as morphine itself. The substance was called heroin after its 'heroic' effects and was promoted from 1898 by German pharmaceutical company Bayer as a non-addictive cure for morphine addiction. Twelve years later heroin was shown to be a far more potent and addictive form of the drug it was supposed to counter. By 1910, as a result of this unintended consequence, the heroin genie was thoroughly uncorked and an epidemic of addiction had taken grip across Europe and America. The situation was so bad that the world's first co-ordinated treaty prohibiting the use of certain narcotics (chiefly opium and cocaine) was signed by thirteen nations at The Hague in January 1912.

Meanwhile an even more serious poppy problem had broken out in the Far East. What China lacked in terms of syringes and heroin was more than compensated for in terms of the sheer volume of opium being smoked by its people in the nineteenth century. The reason for their addiction was the massive increase in supply stimulated by the sowing of poppy plantations in India as a way of re-balancing the unequal trading relationship between the British imperial authorities and China.

Between 1767 and 1850 the quantity of opium exported from India to China is estimated to have increased seventy-fold. By 1799 the Chinese government had became so alarmed at the effects on its people of such a massive increase in supply (and fall in prices) that it banned the import of opium outright, making its possession a capital offence punishable by strangulation. But poppy power, combined with the commercial determination of British merchants, ensured such attempts at prohibition were doomed to failure. The bribery of local officials (themselves addicted) helped further increase the quantities imported, regardless of any imperial ban.

To what extent the social consequences of rampant opium addiction in nineteenth-century China directly contributed to the breakdown of that empire is still a matter of hot debate, but by 1850 China was embroiled in the bloodiest civil war the world has ever seen (as many as thirty million people died in the Taiping Rebellion). Imperial authority was further weakened by the intervention of British forces, which twice invaded China in two Opium Wars (1839–42 and 1856–60).

Western hypocrisy was good news for poppies, which were planted in their millions throughout Bengal in northern India. Cultivation then fanned out eastwards across south-west China, Burma and northern Thailand to feed Asia's addicts. Poppies were aided further in their global propagation by the migration of tens of thousands of starving opium-obsessed Chinese immigrants from their mother country during the Taiping Rebellion (1850–71). Many were taken on by Western colonial powers as 'coolies', since overt slavery was by then mostly illegal. As a result, opium dens and drug-smuggling operations became entrenched in inner-city districts from London to New York, opening up as if in defiant revenge for what had already happened on mainland China. Once poppy plantations sprang up in Mexico and South America during the early twentieth century, the plant's global conquest was all but complete.

By the beginning of the twentieth century, Western governments

Cannabis: Selling, dealing and smoking dope in British India, the source of China's nineteenth-century trauma.

Coca: Papal bull: Leo XIII endorsing the therapeutic effects of cocaine-laced wine.

Poppy: Killer remedy: Mrs Winslow's Soothing Syrup, marketed from 1849 as a medicinal aid for mothers coping with teething children, was laced with one grain of opium (65 mg) per fluid ounce.

Lavender: Perfectly charming perfume as promoted during the 1920s.

Coca: Teething troubles: early users of cocaine were ignorant of possible

Orange: 'An essential source of vitamin C' was how US advertisers in the 1950s persuaded consumers to adopt a juicy new habit.

Tobacco: 'I'm sending Chesterfields to all my friends,' said the actor (and future president) Ronald Reagan in 1952, shortly after links between smoking and cancer were first established.

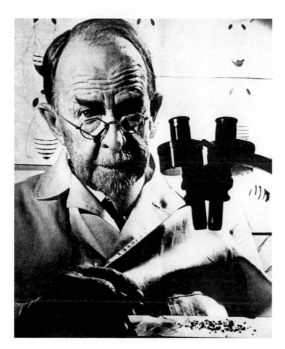

Fruit fly: Thomas Morgan studying fruit flies under a microscope in his US laboratory.

Penicillium: Sir Alexander Fleming pictured examining a Petri dish six years after receiving the Nobel Prize for discovering the antibiotic properties of *Penicillium*.

Grass: An eighteenth-century cutting machine complete with roller to ensure a stripy, symmetrical finish.

Black pepper: Pepper harvesting in southern India as depicted in Marco Polo's widely read thirteenth-century travel journal.

Lotus: Indian goddess Laksmi, wife of Vishnu, rises up on a lotus flower holding her infant Ganesha (illustrated *c.* 1860).

Herpes: Herpes simplex about to latch on to the surface of a soon-to-be-infected cell.

Lactobacillus: Rod-like *Lactobacillus bulgaricus*, the strain found in probiotic bacteria, which secretes lactic acid, inhibiting the growth of pathogens.

were in the grip of an opium-inspired epidemic that still plagues their societies today. Heroin addicts (there are thought to be more than eleven million worldwide) have seldom been able to afford the cost of their habit, estimated to be on average £16,000 a year. Strict prohibition of the drug, especially in America since President Nixon's 'War on Drugs' in the 1970s, means modern addicts and dealers resort to urban violence, thefts and gangland shootings to feed their addiction, events that have become commonplace in even the most salubrious Western cities from Berne to London, and New York to Rome.

There is even a line to be drawn that links poppies with the modern 'War on Terror' triggered by the attacks of 9/11. More than 92 per cent of today's worldwide supply of heroin is derived from the poppy fields of Afghanistan, a terrorist-harbouring country whose economy has been broken for more than 200 years by the repeated invasions of imperialist powers (British, then Soviet, now NATO). Other than growing poppies and trading opium for producing heroin, Afghanistan's economy has little to offer in exchange for the necessities for survival, be they arms or food. In 2007, according to a United Nations drugs report, global heroin production reached its all-time recorded high of 606 metric tonnes.

Penicillium

FAMILY: TRYCHOCOMACEAE
SPECIES: PENICILLIUM CHRYSOGENUM
RANK: 10

A naturally occurring antibiotic that has
transformed modern medicine

AN UNWASHED PETRI dish, an open window and three weeks of unusually cold weather were just a few of the random, chaotic circumstances that apparently led to Alexander Fleming's discovery of the antibiotic properties of a fungus called *Penicillium* (see plate 13, bottom). In the summer of 1928 the Scottish microbiologist had taken a three-week break from his research in the basement laboratory of St Mary's Hospital, London. On his return he casually began to clear up the left-overs of some previous experiments that he should have washed up before he went away. In one dish he noticed that a mould had contaminated a culture of the common infectious bacterium streptococcus. The mould was surrounded by a clear halo. '*That's funny . . .*' said Fleming, gesturing towards his lab assistant, Merlin Pryce.

While luck undoubtedly led to the mould contamination, it was only thanks to Fleming's eagle eyes and vast experience as a microbiologist that the enormous significance of that halo effect was determined. Fleming realised that the mould must be secreting a compound that was killing off the bacteria. Fleming's discovery was the equivalent of biological dynamite because so many infections and diseases that afflict animal life (including humans) are caused by a group of bacteria classified as 'Gram-positive' – like those streptococci in the Petri dish. Fleming went on to write a paper about the antibacterial properties of the fungus, which he identi-fied as *Penicillium notatum*. The antibiotic compound it produced he called Penicillin G. Fleming was *not* responsible, however, for turning the discovery of *Penicillium*'s powerful antibiotic properties into something that had worldwide impact. That's why, when Nobel prizes were awarded in 1945, he shared it with two other people – Australian scientist Howard Florey (1898–1968) and German-born biochemist Ernst Chain (1906–79).

Florey and Chain were part of a team of microbial researchers in Oxford, England. By chance Chain picked up the scientific paper written by Fleming in the 1930s that proposed using the *Penicillium* fungus as an antibiotic. Wounds sustained during World War II frequently became so badly infected that amputations were often the only way to save a victim's life. Even then, success was a hit-and-miss affair. In those days infection was feared as much as cancer is today.

On Saturday 25 May 1940 Howard and Chain conducted an experiment on eight mice injected with a lethal dose of streptococci bacteria. Four were treated with penicillin and, as controls, the other four were left untreated. When Florey and Chain arrived at their lab the next morning they found, alongside the four dead untreated mice, the four treated mice restored to perfect health. Human trials began in earnest. One of the main stumbling blocks was how to grow sufficient quantities of the fungus to administer a big enough dose to treat a human. One of the less well known (and less salu-brious) production techniques initially used by the team was the utilization of hospital bed pans to grow mould. Liquid containing penicillin was drained from beneath the growing mould and filtered through parachute silk into milk bottles on nearby bookshelves. But such makeshift techniques could never produce sufficient quantities to make much impact on the masses of wounded soldiers needing treatment during the war.

It was another species – a most unlikely find – that came to the rescue thanks to some speedy transatlantic collaboration. An American cantaloupe, a type of melon originally taken to the New World by Christopher Columbus on his second voyage in 1494, was found to be the best medium for growing a *Penicillium* species called *chrysogenum* that produced up to 200 times more Penicillin G compound than Fleming's *P. notatum*. This surprising mould-growing habitat was discovered by Mary Hunt (also known as 'Mouldy Mary'). During 1943 a series of experiments that involved shooting X-rays through Mary's mould produced a mutant *Penicillium* variety that further increased the quantity of antibiotic compound to 1,000 times more than that in Fleming's original culture. Giant aerated 87,000-litre metal tanks (instead of milk bottles) were deployed to grow the mould so that by 1945 enough penicillin was being produced to treat as many as *seven million* patients a year.

For the genes of the fungus genus *Penicillium*, the partnership with modern human beings has been a propitious one. Antibiotic sales today are estimated at more than £6 billion a year. Infections that encounter penicillin suffer the bacterial equivalent of a holocaust, especially in domesticated animals regularly injected with antibiotics. A constant state of war now exists between different forms of artificially selected antibiotics and the super-fast evolutionary footwork of naturally occurring bacteria. With their short generation times, these microbes evolve at warp speed. So far humans and their fungal partners have generally had the upper hand – although thanks to the frequent overuse of antibiotics, resistant bacterial infections (such as MRSA) are reaching epidemic proportions.

The impact of penicillin on people at a personal level over the last sixty years has generally been one of enormous relief as millions of lives have been saved from otherwise fatal infections. On a global scale the development of penicillin may be more of a double-edged sword. The increase in the human population afforded by its widespread use has been estimated at more than 200 million individuals

– not quite in the league of artificial fertilizers or Norman Borlaug's Norin 10 (see page 202) but very substantial nonetheless.

Florey, regarded as one of Australia's most important scientists, was personally extremely conscious about the global impact of the wonder drug he helped bring to market. '*I am now accused,*' he said in 1967, '*of being partly responsible for the population explosion . . . One of the most devastating things that the world has got to face for the rest of this century.*' Florey became so sensitive about the issue that he dedicated much of his later research into developing improved methods of contraception.

Cinchona

> **FAMILY: RUBIACEAE**
> **SPECIES: CINCHONA OFFICINALIS**
> **RANK: 84**

A tree that provides humans with their most
effective defence against malaria

PLANTS ARE VICTIMS of human discrimination and prejudice every bit as much as people. Suspicion is frequently propagated by religious tradition, as with Mohammed's intolerance of plants that ferment into alcohol, or the Christian Church's vilification of fungi and herbs such as cannabis, hemlock or belladonna as ingredients in 'witches' brews'. Foods regarded as staples today, such as the highly nutritious potato, suffered hundreds of years of social stigma before they were accepted as good to eat in Europe (see page 210). Even when a plant offers an apparently miraculous cure from death by uncontrollable fever, prejudice can still intervene.

In 1658 Oliver Cromwell, a politician and soldier who usurped power from the British monarchy during the English Civil War (1642–51), lay on his deathbed having contracted what many historians believe was malaria while campaigning in Ireland. A Venetian medic offered him a possible cure – a solution based on the powdered bark of a tree native to South America. Despite his high fever, Cromwell dismissed the potion out of hand, fearing it was part of an elaborate popish plot. Instead he was bled, a procedure that inflamed his condition still further, and he died soon afterwards. We are therefore left to ponder how differently Britain's history might have turned out had Cromwell cast aside his Protestant prejudice.

On Drugs

No one knows for how long the active ingredient quinine, found in the powered bark of the cinchona tree, has been used by people as a defence against the often fatal effects of the malarial parasite *Plasmodium falciparum* (see page 131). There is no explicit evidence of its use before Spanish invaders arrived in the Americas in the late fifteenth century, perhaps because, like smallpox, malaria travelled with them from Europe and Asia. But once this disease – spread by bloodsucking mosquitoes – broke out in South America, the indigenous people soon put their prodigious knowledge of herbal medicine to work.

By the 1620s Jesuit missionaries working in the forests of the Andes had learned of the healing effects of the powdered cinchona bark from natives. Its first noted success was in 1630, when the governor of Loxa, a thriving Peruvian city, drank an infusion made from cinchona bark and was cured of a life-threatening fever. Jesuit priests working in South America were the main conduit for the diffusion of the miraculous bark back to Europe, where malaria sufferers in Italy and Spain made similarly spectacular recoveries. It is this association that led to decades of argument between Catholics and Protestants about the pros and cons of using a papist remedy known as 'Jesuit's bark'.

It took a medic with exceptional cunning and entrepreneurial skills to break down this prejudice and finally to convince even Protestant nations that here indeed was an effective cure for a most pernicious disease. Robert Talbor, a Cambridge physician born in 1642, became famous for concocting a secret remedy for high fever. Aware of popular misgivings against the use of Jesuit's bark in England, Talbor wrote a book in which he actively discouraged people from taking this remedy: '. . . *beware of all palliative cures and especially of that known by the name Jesuit's powder, as it is given by unskilful hands*'.

Talbor's fever-busting special remedy became so famous that King Charles II of England (ruled 1660–85) appointed him Royal

Physician in 1672 and knighted him in 1678. The following year Charles fell ill with fever but was cured by Talbor's remedy, as was the son of King Louis XIV, who paid Talbor 3,000 gold crowns out of gratitude for saving the dauphin's life. When Talbor died in 1681, his secret remedy was finally revealed – a solution of opium, alcohol, lemon juice and . . . guess what? Powdered bark from the cinchona tree.

European efforts to cultivate the plant were long frustrated, however. When in 1735 French naturalist and epic explorer Charles Marie de la Condamine travelled all the way down the Amazon river (see also rubber tree, page 261), he took the trouble to collect what he believed would be a vastly valuable cargo of cinchona seedlings, planted in boxes of earth, which he carefully nurtured during his eight-month journey through jungle, swamps and rapids. Just as he came within reach of the coast – in sight of the ship that would take him and his precious cargo back to Paris – his small boat was swamped by a wave and the cinchona plants washed away.

Subsequent 'seed-napping' efforts by other Europeans failed to achieve successful cultivation of the cinchona tree outside South America until 1865, when an Englishman called Charles Ledger finally managed to break the Peruvian monopoly. He obtained seven kilograms of seeds from his native servant, Manuel Incra Mamani, for about twenty dollars. Ledger's seeds were sold to Dutch colonialists, who established plantations in Java on the other side of the world. British and German colonialists followed suit, extending the tree's reach into India and Ceylon (now Sri Lanka).

French scientists successfully isolated the active compounds in the cinchona tree – now known as quinine – in 1822. Along with a big increase in the worldwide availability of cinchona bark, thanks to the new plantations in Asia, the battle between humans and the *Plasmodium*-carrying mosquito began to tip in humanity's favour. The mass administration of quinine in the late nineteenth century made it possible for Europeans to colonize malaria-infested regions of equatorial Africa. The 'Scramble for Africa', sanctioned by the Congress of Berlin in 1884–5, therefore owed at least as much to the global spread of quinine as it did to the forces of European avarice, industrialization, steam power or mass-produced armaments.

Malaria, for which there is no prospect of a vaccine, is still a big killer today – especially in Africa. Attempts at producing an artificial alternative to natural quinine have been only partially successful. *Plasmodium* parasites have successfully evolved resistance to chemical alternatives such as chloroquine, daraprim and mefloquine, leaving natural quinine by far the most effective treatment. The cinchona tree has triumphed over human prejudice in the last 500 years, spreading itself with human help all over the world. In the process it has sustained and increased populations thanks to the remedial powers of its bark – a vital component in humanity's medical kitbag, which we look set to rely on well into the future.

12

On Companionship

How some animals have prospered by being good company but have recently provoked a moral dilemma.

Dog

FAMILY: CANIDAE
SPECIES: CANIS LUPUS FAMILIARIS
RANK: 42

*A creature that has succeeded in becoming
mankind's best friend*

ALTHOUGH THE EXACT sequence of events that led to
the cohabitation of dogs and humans is still hotly disputed, it is
now widely accepted that today's domestic dogs all belong to one
subspecies of the grey wolf (*Canis lupus*). Genetic diversity suggests
east Asia as a location for their initial domestication and there is
speculation that wolf puppies were initially taken in and looked after
by humans so they could be groomed to pull sleds in what was then
still an extremely icy world.

Asian humans who migrated across the frozen Bering Straits to
populate North America approximately 13,000 years ago may have
succeeded in the perilous crossing only because their supplies were
hauled over the ice by the ancestors of today's husky dogs. At about
the same time, canine company was so precious to early agriculturalists
living in the Near East that human graves have been found with the
remains of beloved dogs lying alongside the bodies of their owners.

Countless generations of breeding by humans have resulted in
the evolution of a number of different races of dogs. Some have
been groomed for herding sheep (collies), others for catching vermin
(terriers), transporting goods (huskies) or fetching game (retrievers
and pointers). By medieval times the idea that a dog was man's best
friend was well established. Effigies of wealthy European aristocrats

and soldiers are seldom seen without a dog at their feet, a symbol of chivalry, courage and fidelity.

Loyalty is commonly cited as one reason why domestic dogs have so successfully established themselves in a world where the human population has grown from approximately five million 12,000 years ago to nearly seven billion today. Dog numbers have now reached an estimated *400 million* individuals worldwide (compared to wild wolves, which although still one of the most widely distributed animal species, have dwindled to approximately 200,000).

A prodigious ability to learn is another attribute that has helped domestic dogs succeed in the human world. There is no established or standard method for measuring the intelligence of dogs (or humans, for that matter) but their capacity to comprehend, respond to and obey human demands is well established. A recent experiment conducted by animal psychologist Juliane Kaminski has shown how Rico, a border collie, can understand a vocabulary of more than 200 labels for different items, including being able to interpret commands such as 'Fetch a sock' and give it to a particular person. A dog's capacity for learning about the human world and with it the commands of human beings has been, like that of horses (see page 246), instrumental in their success as a key domestic species.

Also like horses, dogs are extremely sensitive. With a listening range of between 40 hertz and 60,000 hertz, these creatures are far superior to humans in their hearing capacity at both ends of the audio spectrum. They use as many as eighteen muscles in each of their ears to focus sounds, enabling them to detect and pinpoint noises four times farther away than those heard by humans. While their vision is at least as good as that of humans (although they do not have the capacity to perceive as many colours) it is their extraordinary sense of smell that makes all human–canine comparisons begin to look ridiculous. Relative to brain size, the olfactory cortex in a dog's brain is forty times larger than the equivalent processing centre for smells in humans, and while the number of smell receptors that feed information into a dog's brain totals 220 million, the equivalent in humans is approximately five million, making a dog's sense of smell vastly more sensitive.

This prodigious olfactory sense is derived from the habits of the domestic dog's wild grey wolf ancestors, creatures that define their hunting territories by leaving traces of urine as scent marks. Whereas humans identify each other through visual recognition, dogs, like wolves, secrete individual pheromones through glands in the base of their tails, between their toes, in the genitals and skin. Individual identification via smells explains why dogs, when they meet each other for the first time, typically begin their acquaintance by sniffing each other's bottoms. Throughout recorded history humans have compensated for their own olfactory inadequacy by employing dogs for everything from hunting rats and foxes to modern law enforcement (detecting drugs) and even in the search for the human survivors buried in rubble caused by disasters such as earthquakes and terrorist attacks.

No less essential to the enormous success of dogs is their extraordinary genetic malleability. Although several breeds had been successfully created over thousands of years to suit specific human demands, it is only in the last 150 years that most of today's 300-plus pedigree breeds have come into existence. From *c.* 1850 some pet owners began to experiment by breeding dogs into particular shapes and forms for the sole purpose of attracting attention or participating in competitions. Lapdogs became a social accessory for Victorian ladies while bulldogs became a symbol of British valour in war. Dog shows such as Crufts (founded in 1886) and organizations such as the Kennel Club (which took over Crufts in 1938) were founded in Britain but have since become worldwide catalysts for modern canine eugenics – selective breeding for the sake of achieving desirable traits in future generations. New breeds have been created through the generations by culling or sterilizing animals that do not conform to certain breeding rules (such as size, fluffiness, butchness and so on) while owners ensure the survival of those that do.

It was largely thanks to the huge success of Victorian dog breeders from *c.* 1850 that Charles Darwin's cousin Francis Galton (1822–1911) was inspired to propose that similar breeding techniques be used on humans. Galton's *Inquiries into Human Faculty and its Development* (published 1883) suggested human marriages should in future be based on identifying people with desirable traits (such as intelligence and honesty) and that incentives should be created for these people to marry and have children. *Kantsaywhere* was a novel Galton wrote about a utopian society where a religious order was responsible for breeding smarter, fitter humans. Only fragments of this work survived after Galton's niece burned much of it, so offended was she by its love scenes.

Galton coined the term eugenics for the science of selective human breeding. It rapidly became a popular political topic in the United States as a possible solution to racial divisions (as proposed by writers such as Madison Grant in 1916). Eugenics was later adopted with devastating fervour in Nazi Germany. Adolf Hitler attempted to revive a mythological master race of humans (Aryans) through the culling of 'inferior' beings (mostly Jews) and the enforced sterilization of thousands of other 'undesirables'. Hitler's eugenic ideals were supported by a perverse belief in animal rights in which wolves were second to human Aryans in the natural pecking order of life. Other human races, such as the Jews, were subordinate to dogs. Historians are left to ponder to what extent Hitler's ideals were influenced by the love he had for his German shepherd, Blondi, which his secretary gave him in 1941.

The impact of domestic dogs on human culture therefore reaches far beyond their wide-ranging usefulness to mankind. From the rise of the animal rights movement to the dark experiments of modern eugenics, these are creatures that have been, like horses, true protagonists in the turbulent history of human civilizations.

Cat

> **FAMILY: FELIDAE**
> **SPECIES: FELIS SILVESTRIS CATUS**
> **RANK: 46**

An unlikely bargain struck between agricultural
people and a small carnivore that gave rise
to the world's most populous pet

FEW ANIMALS IN the history of civilization have been afforded
as much hospitality as the domestic cat. In ancient Egypt they were
thought to be divine. The cat-goddess Bast had her own picturesque
shrine and cats were, like humans, frequently embalmed so they
could return to their earthly bodies in the afterlife. In 1888 a farmer
found an ancient burial site along the banks of the river Nile that
contained *80,000* mummified cats. According to Greek historian
Herodotus, who visited Egypt in 450 BC, cats were so venerated that
when one died its owners were '*plunged into deep grief*' and would
'*shave off their eyebrows*' to signify the loss.

In Islam the Prophet Mohammed was a renowned lover of cats
– which he regarded as clean animals, unlike dogs. Tradition has it
that during one prayer time Mohammed found his pet cat Muezza
asleep on the sleeve of his overcoat. The prophet cut off the sleeve
so as not to disturb the cat's slumber. Mohammed's legendary love
for his feline friend meant that a high regard for cats has continued to run throughout Islamic cultures. In Japan one cat is said to
have waved its paw at a passing nobleman. As the man turned to
approach it, a bolt of lightning struck the ground where he had been
standing. The beckoning cat that saved this unknown aristocrat's

life has since become Japan's most widely recognized sign of good fortune. Even today *Maneki Neko* ('beckoning cat') dolls welcome people into shops and homes throughout the country. In Europe cats became common household creatures following the Roman conquest of Egypt in 31 BC. They were transported by early modern European maritime explorers to the Americas and Australia, where they previously had no native presence. The consequence of such cultural alliances and global transportation is that domestic cats have recently eclipsed even dogs in terms of their population, now estimated at *600 million* individuals worldwide.

The extraordinary success of cats presents a conundrum because these are not creatures that have been trained or domesticated by humans to provide any obvious kind of service. They do not transport things – like dogs, horses and camels do – nor do they provide humans with a source of food, drink or clothes – like cows, sheep and pigs. Even as a badge of social status, cats have mostly failed to impress other people in the same way as a pedigree dog or thoroughbred racehorse. What is it, then, that accounts for the domestic cat's present and historic success? How come hundreds of millions of people all over the world and throughout the history of civilizations have welcomed cats into their homes?

The earliest archaeological evidence for domestic cats goes back about 9,500 years to graves in Cyprus, where cat remains have been found buried alongside those of humans. Recent genetic analysis of 979 domestic cats suggests they are all a subspecies of the wildcat (*Felis silvestris*), which itself evolved from a common ancestor about 230,000 years ago. Today's domestic cats are thought to have derived from as few as five *F. silvestris* females that originated in the Near East at about the time humans began experimenting with a lifestyle based on farming approximately 12,000 years ago.

Unlike other domesticated farm animals, cats do not seem to have been 'selected' by humans for any specific purpose. In fact, cats are just as likely to have chosen to live with humans in a process known as 'self-domestication'. Experts believe that once people began farming in the Near East, it was a survival advantage for cats to become less fearful of humans because stores of grain and other produce were a constant magnet for the animal's favourite prey – rats and mice. A good life could be had – a life of fun, playing, teasing and killing rodents – in the granaries and barnyards of human farmers.

One reason why humans initially welcomed these exotic-looking creatures into their living quarters may also be due to a curious ancient genetic accident. Some time between the origin of *F. silvestris* and their much more recent domestication, cats lost the use of taste buds that detect sweetness (a mutation or viral infection, perhaps). As a result, domestic cats have evolved into 'obligate' carnivores, creatures that feed exclusively on flesh. That cats are unable to eat fruit or seeds must have given early human farmers sufficient comfort that these creatures would not harm precious stores of agricultural produce. Instead, they would focus their advanced, stealthy hunting skills on whatever invading vermin came into range. This human–cat self-domesticating symbiosis seems to have begun in Egypt – the bread basket of the ancient world. Cats tolerated humans, becoming tame and friendly towards them, while humans tolerated cats as the chief guardians of their food.

It is remarkable that humans have become so close to cats for, unlike domestic dogs (see page 336), they are mostly solitary self-sufficient creatures. While cats sometimes live in colonies (today as many as 300,000 cats roam freely in Rome), they almost always hunt alone and fiercely defend their own territory. Such behaviours are usually regarded as the antithesis of those characteristics required to live with humans. But cats compensate by behaving in ways that are curiously reminiscent of human habits, perhaps explaining why they have got along so well with people over the last 12,000 years. Cats, like people, hunt other animals for pleasure (we call it sport), triumphantly bringing trophies of dead birds or mice back into their homes to show off to their human foster parents. Also, like many

On Companionship

people, cats have a penchant for psychoactive drugs – their favourite being catnip (*Nepata cataria*), a common herb that they chew, sending them into a euphoric state for up to two hours. Nor do cats, like humans, respond well to being enslaved. They prefer a degree of personal freedom and to live semi-independent lives, coming and going as they please, sometimes indoors, sometimes outside, where like entrepreneurs or nation-states they establish and defend their own territorial patch.

The question as to whether cats bring good or bad luck has fascinated humans throughout recorded history. While Egyptian, Islamic and Japanese cultures have generally venerated cats, in medieval Europe their image took a hammering from their association with Gypsies and witchcraft. Cats were frequently cast as accomplices of the devil because of their solitary habits as night-time hunters, which often seemed coolly detached from human dependence.

By Victorian times, as the Age of Reason prevailed over the age of witchcraft, cats recovered from such satanic associations. Breeding programmes were established to improve their looks, distancing most varieties from any association with a witch's black cat. From the mid-nineteenth century, cats became much-adored companions for everyone from pet-loving children to lonely adults such as Victorian author and poet Thomas Hardy, who famously commemorated the death of his beloved pet cat in his poem 'Last Words to a Dumb Friend' (1904):

> *Never another pet for me!*
> *Let your place all vacant be;*
> *Better blankness day by day*
> *Than companion torn away.*

Rabbit

FAMILY: LEPORIDAE
SPECIES: ORYCTOLAGUS CUNICULUS
RANK: 28

*A domesticated species that has recently been decimated
as its population soared out of control*

RABBITS BELONG TO a family of rodent-like mammals called
the Leporidae, which also includes hares and pikas. Although more
than twenty species of wild rabbit are known, domestic rabbits all
belong to a single species – *Oryctolagus cuniculus* – native to the
Iberian peninsula.

For most of their history, domesticated rabbits have been kept as
food, not pets. The Romans – notorious for spreading every species
from grapes to cats all over their empire – brought rabbits from
Spain to Italy about 2,200 years ago, eyeing these fast-reproducing
herbivores as a good source of easy-to-farm meat. Multi-tiered
amphitheatres, triple-decker aqueducts and long straight roads
are what people generally perceive as the most enduring impacts
of this once-mighty European civilization. But in many parts of
the modern world the consequences of the initial transportation
of rabbits across Europe far outstrip these remaining marvels of
ancient engineering.

The burrow-making habits of European rabbits explain in large
part why this species has been so successful compared with hares
and other types of rabbit (like cotton-bobtails) that live only above
ground. Females dig tunnels and underground chambers where they

give birth to their young. The young bunnies benefit from the sheltered environment in the first few weeks of their lives. The brilliant burrowing antics of European rabbits also turned them into expert escapologists. Not only are they able to jump high walls using their strong hind legs but they can tunnel deep beneath fences and other artificial enclosures. Therefore, wherever rabbits have been transported by humans for the sake of farming, they have also managed to escape into the wild to seed feral populations.

A second success factor is their prodigious rate of reproduction. Unlike in other mammals, ovulation in female European rabbits is triggered by copulation, which means that instant fertilization is almost guaranteed. With a gestation period of just thirty-one days and litter sizes of up to twelve bunnies at a time, rabbits represent one of nature's most prolific reproductive systems (see also chickens, page 232).

From Italy, rabbits spread throughout medieval Europe, reaching Britain shortly after the Norman Conquest. Outdoor warrens were established by farmers all over the country, eager to cash in on the sale of rabbit meat and fur. Harvesting was accomplished by sealing burrow entrances with nets and putting a predatory animal such as a ferret into the maze of tunnels to flush the rabbits out (lobsters and crabs were used closer to the coast). Although wild populations in Europe grew prodigiously as more rabbits escaped from warrens, natural predators such as foxes, stoats and lynxes kept their numbers reasonably in check. By the nineteenth century, shooting rabbits became a popular British pastime too, as population control became a competitive blood sport.

During the nineteenth century, however, the transportation of rabbits to other habitats proved environmentally catastrophic. Mirroring the ecocide inflicted on parts of Europe and Asia by the introduction of species such as the Australian eucalyptus (see page 271), European rabbits inflicted unquantifiable damage to ecosystems, especially where there were no predatory creatures to control their numbers. Islands from Lundy to the Falklands were turned into rabbit-infested ruins because European sailors had treated these places as breeding grounds – since rabbits cannot swim they could not escape. The most notorious introduction took place in October 1859 when homesick Australian settler Thomas Austin asked his English nephew to send over 24 breeding rabbits as game on his 12,000-acre estate at Winchelsea, near Geelong in the southern Australian state of Victoria. Soon after the creatures arrived Austin released them into the wild, declaring to his friends that '*the introduction of a few rabbits could do little harm and might provide a touch of home, in addition to a spot of hunting*'.

Without the threat of natural predators, and with their extraordinary reproductive prowess, rabbits progressed across Australia at a rate of about 110 kilometres per year. By 1865 Austin had killed 20,000 on his own estate alone. By 1883 the government of New South Wales had spent £361,492 on killing off 7.8 million rabbits. By 1950 the rabbit population of Australia had reached *600 million*. It was the fastest spread of any mammal ever recorded anywhere in the world.

The effect on the country's habitat was devastating. Soil erosion (from burrows), deforestation (from eating young plants and seedlings) and a decline in native creatures (such as the bilby, a rabbit-like marsupial) were direct consequences. Crop destruction was of greater concern to humans, however. A Royal Commission was established in 1901 to try to find a solution, following which a fence was erected in 1907 – the longest in the world – stretching for more than 1,700 kilometres across western Australia in a bid to keep the invading rabbits at bay.

Of course, it never really worked. Fences, traps, shooting, poisons – none of them provided any significant impediment to the spectacular burrowing and reproductive success of escaped domesticated European rabbits living in a world bereft of natural predators. Ironically, by this time, domestic rabbits back in Europe – the source of all the trouble – had become one of Europe's most treasured household pets. The Victorian fancy for household animals led to the establishment of as many as 150 different cosmetic breeds based on the colours of their coats and the length of their ears. Since 1975 pet rabbits have even been bred for competitive sports – Sweden and Britain lead the world in rabbit 'show-jumping' events.

Until 1950 the planet Earth had looked kindly on the genes of *Oryctolagus cuniculus*. In the southern hemisphere the species had successfully overwhelmed ecosystems while in its indigenous northern homeland it was given additional protection as a popular household pet. As the Western pet-keeping craze developed, even the traditional habits of eating rabbit meat and wearing its fur became taboo in countries such as Britain. But neither nature nor man guarantees a species prosperity for ever. Frank Fenner (born 1914), an Australian virologist, offered humans their first real chance of striking back at the rabbit's soar-away success. Fenner discovered that the *Myxoma* virus infected *O. cuniculus* specifically, causing the disease myxomatosis, which had an extraordinary 99 per cent mortality rate. Fenner proved the virus was perfectly safe to humans by injecting himself with it before it was released into the Australian wild in the summer of 1950.

The effect was devastating. The disease quickly spread through the feral population via flea and mosquito bites and within two years more than *500 million* Australian rabbits had been killed. As the rabbit population declined, so the mortality rate decreased (a viral evolutionary tactic to ensure numbers do not reduce so much that the virus no longer has a host population in which to thrive). Of course, some rabbits were naturally immune (thanks to the insurance system of sexual reproduction, see page 359) so that there was no real risk of total extinction. The disease spread to Europe in 1952, thanks to its illegal release in France. Similar devastation occurred, this time destroying so many rabbits that other wild animals that depend on them for food (such as the Iberian lynx and Spanish imperial eagle) are today threatened with extinction.

What on Earth Evolved? . . . in Brief

The history of domesticated rabbits shows how chaotic the relationship between man and other species can be, especially when rational human minds try to harness a species for their own ends. Loved as pets, loathed as pests and depended on as a viable source of meat and fur, rabbits provoke attitudes in man that are riddled with contradictions. However, rabbits can claim to have proved capable of surviving even the most devastating turns of fortune.

Hamster

*Unsung wartime heroes that have since become
one of the world's most popular pets*

WHEN SAUL ADLER challenged his Jewish zoologist friend
Israel Aharoni to go on a hunting trip to the Syrian desert in 1930,
little could he have imagined how extraordinary the consequences
of the expedition would turn out to be. Adler, a Russian-born
researcher into tropical diseases, was desperate to find a cure for the
parasitic disease leishmaniasis, spread by bloodsucking sand flies.
His research depended on the discovery of a new species of desert
hamster.

Records concerning this sometimes fatal illness stretch back
at least 2,500 years. Stone tablets found by archaeologist Austin
Layard in the royal palace of Assyrian King Ashurbanipal (died 627
BC) describe the symptoms almost perfectly, which include blistering sores on the skin. Muslim medic Avicenna (AD 980–1037) also
sought a cure in tenth-century Baghdad. Blotchy people depicted
on pots made in pre-Incan South America show the global reach
of this protozoan parasite. Leishmaniasis is contracted by hundreds
of thousands of people each year in areas as far flung as Colombia
and Afghanistan. After invading the body in a sand fly's saliva, the
Leishmania parasite spreads through the bloodstream infecting a
number of different tissues, eventually causing severe inflammation
of the liver and spleen. Skin sores usually develop weeks or sometimes

months after initial infection. If left untreated, the disease leads to a particularly unpleasant death.

Drugs that can kill the parasite have been developed over the last eighty years mostly thanks to that Syrian expedition undertaken by Aharoni. Adler had been using Chinese hamsters in his Jerusalem laboratory to carry out research into a drug cure for leishmaniasis. Through repeated infections these hamsters were immune and no longer fell victim to the disease. What Adler needed was a similarly small, fast-breeding hamster species that was just as content to live alone in confinement so he could continue with his research. But it must be a species that was not immune.

The unsung heroes of the medical breakthrough that finally helped find a cure for the ancient disease leishmaniasis were a single wild female golden hamster and her litter of eleven babies, living in a burrow in the desert near Aleppo, discovered by chance by Aharoni. Here was a new species, never before exposed to the disease, which researchers could use to continue with their experimentation. On the way back to Jerusalem the mother died, leaving Aharoni the task of feeding and caring for the babies himself. By the time he got back to Adler's lab, four had survived. Recent DNA analysis has confirmed that it is from these four individuals that all the golden hamsters kept in captivity today, and sold in pet shops throughout the world, are descended.

In Sicily during World War II, large numbers of Allied troops contracted leishmaniasis from sand fly bites and were at serious risk of death. A cure was desperately sought. In 1943 British disease expert Leonard Goodwin (1915–2008) contacted researchers in Israel for a consignment of Syrian golden hamsters on which to test some new drugs. Thanks to their lack of immunity to the *Leishmania* parasite, Goodwin was able to calculate the correct dosage for administering the drugs to humans without risking death from side-effects of the treatment. Within a year the drug pentostam was approved and used to treat the troops. After the war Goodwin donated the research animals in his laboratory off Euston Road, London, to some pet fanciers. The fast-breeding easy-to-keep solitary handsome golden-coated creatures quickly became favourite first-time pets for children. Selective breeding from this original colony has introduced a number of other breeds, with different-coloured coats, adding to the range and choice of hamsters available in pet stores today.

The present-day impact of these creatures, countless millions of which are now caged in homes all over the world, is less easy to quantify than the heroic contributions of their wartime ancestors. Often these tame rodents are an urban child's first close-contact experience with an animal. Such experiences can have profound effects – Goodwin himself was inspired to pursue a career in biology after his grandfather, who was a gamekeeper, took him out of London on nature holidays to Rutland in the East Midlands. In the wild golden hamsters are rated as a 'vulnerable' species because of habitat loss. But, thanks to their chance discovery by a man on a mission near Aleppo in 1930, the Syrian variety *Mesocricetus auratus* has become one of the most successful domestic species in the modern world.

13

On Beauty

How good looks, strong smells and powerful flavours in some species have proved irresistible to humans.

Rose

> **FAMILY: ROSACEAE**
> **GENUS: ROSA**
> **RANK: 62**

An ancient symbol of love and romance that has
recently become unsustainably successful

LAKE NAIVASHA IS a 130-square-kilometre stretch of African water that has been a paradise for wildlife throughout the history of humanity. Located in the heart of Kenya, one hundred kilometres north of Nairobi, the lake straddles the Great Rift Valley, a region richly endowed with volcanic soils where some of the earliest hominid fossils were discovered in the early 1970s. An abundance of fish, lions, antelope, leopards and hippos thrived in this habitat alongside more than 350 recorded species of wild birds. Until the 1970s about 7,000 people lived around the lake's shores, many of them belonging to the Maasai, settlers from the expansion of west African Bantu herds-people shortly after their cattle developed immunity to a disease carried by the tsetse fly (see page 138).

But in the last thirty years the region has been transformed as if someone had struck gold. More than 300,000 people now live or work on newly established farms around the lake's shores. Aside from the frequent use of harmful pesticides (such as DDT, see page 95) that damage the area's ecosystem, the lake's water level has fallen three metres below minimum safety levels. If current trends continue, says Dr David Harper, a biologist at the University of Leicester, the lake will become a '*turgid, smelly pond with impoverished communities eeking out a living along the bare shores'*.

The reason for this impending ecological collapse defies common sense. Kenya, one of the driest places in the world, is exporting vast quantities of its precious water to some of the planet's wettest countries in Europe. The main reason for this is, in a word, the rose – a flower that for thousands of years has been admired for its beauty and fragrance.

Water accounts for about 90 per cent of the mass of a cut flower. Roses are cultivated in their millions on the commercial flower farms that have sprung up around Lake Naivasha over the last thirty years. These flowers are exported by airfreight to Europe in ready-tied well-watered bunches, where they are sold by florists and supermarkets. The Kenyan flower industry, commercially viable thanks to plentiful supplies of cheap labour, is responsible for generating precious income amounting to some $350 million a year, 74 per cent of which comes from growing and selling roses. One farm, owned by an Indian enterprise, produces *600 million* roses a year. Every stem exports vital water out of arid Africa 365 days a year into rain-rich Europe – and all because of the human obsession with these fragrant mythical flowers.

Roses are loved by humans because they provide both a visual feast and a fragrant bouquet. One of the basic ingredients for perfumes sold throughout the world is oil made from damask roses, which originated in Persia. They were brought to Europe by crusaders in the thirteenth century, along with Arabic oil-distilling techniques, and the practice of growing roses was eventually established in Bulgaria and the Netherlands thanks to their well-watered soils. Today Bulgaria's Rose Valley is the world's largest rose-growing region, producing oils for the perfume industry. Its 300-year-old rose-growing tradition is centred around the town of Kazanlak. Thanks to its mountainous terrain and wet climate, there are no ecological concerns about growing roses here. Traditions live long.

Their agricultural methods have so far resisted the ecologically damaging practices of modern intensive farming and on the first weekend in June the locals, dressed in traditional costumes, flock to the annual flower festival. The highlight of the event is a beauty pageant in which one girl, chosen from among many, is crowned Queen of the Roses.

Roses have been a powerful symbol of beauty from the ancient civilizations of Persia, Greece and Rome to today's annual Valentine's Day celebration. On average 10,000 metric tonnes of roses are sold in the UK each year as gifts to loved ones on Valentine's and Mother's Day.

Five petals that unfurl into myriad shapes and colours, some highly scented, some not, underlie the other great secret of the rose's enduring success. These are flowers, like oak trees (see page 142), that so easily hybridize that the broader term 'genus' is really more appropriate than the narrower 'species'. Botanical enthusiasts are able to cultivate flowers that suit almost every nuance of human fancy in colour, shape, form or fragrance. Once a desirable variety has been grown, horticulturalists can replicate its genetic formula *ad infinitum* by grafting (cloning) a parent plant on to the root stock of another plant (see apple, page 359). When varieties of roses that bloom throughout the year came across from China to Europe in the 1800s, they triggered a craze for horticultural experimentation with rose genes that has never lapsed. Such genetic pliability has provided botanists with a broad palette of possibilities with which to adapt to changes in human taste, style and fashion.

The cost of sustaining our fondness for these flowers doesn't just threaten the ecology of countries like Kenya. As the lake's waters dry up, large cut-flower producers are busily buying up land in other cheap-labour markets, such as in Ethiopia, incentivized by a ten-year tax break from the government of a country that is also chronically short of water. Will the appeal of the rose prove so enduring that African farmers will continue to cultivate them and European consumers continue to purchase them, despite the ecological consequences? The problem of short-term gain over long-term pain seems to be as perennial as the rose itself.

Apple

> FAMILY: ROSACEAE
> GENUS: MALUS DOMESTICA
> RANK: 53

How the genes of a highly nutritious fruit with greater than average diversity were reduced by humanity's desire for visual perfection

JOHN CHAPMAN IS one of America's oldest and most famous iconic heroes. The eighteenth-century frontiersman (1774–1845) earned the nickname 'Johnny Appleseed' because of his obsession with planting apple trees. He would pre-empt the arrival of new settlements by establishing fenced-in apple plantations before moving on to a new area, leaving his orchards under the stewardship of the locals. Every few years he would return to pick up a payment made in kind or as rent. Therefore, every time settlers pushed west, Johnny, the nomadic holy-man-cum-horticulturalist, ensured they were welcomed by a dependable supply of rich juicy fruit.

Apples were everything to North American prospectors following their successful war of independence from Britain (1775–83). Sweetness, a flavour humans have enjoyed throughout history, largely came from fruit. Sugarcane was rare, and honey supplies were limited. It was mostly thanks to Johnny, the thoughtful apple grower, that essential fruit supplies were available to the settlers of Ohio and Indiana. It wasn't just his curious nomadic lifestyle (he never wore shoes) or his clear foresight that make Chapman's story intriguing. More remarkable was his subversive style of horticulture, which embraced, rather than competed with, the evolutionary systems of the natural world: he *always* planted his orchards from seed.

Apples are remarkable for their huge genetic diversity. Buried inside the apple's core are up to a dozen seeds. Grow trees from all twelve seeds and each one will have remarkably different characteristics. This is the effect of sexual reproduction working at its explicit best. Regardless of where apple seeds are carried by their target transporters – mammals – such diversity ensures that at least one or two are able to respond favourably to the prevailing environmental conditions. This statistical bet has helped apple trees adapt to habitats far removed from their native origin in central Asia.

But by the time Chapman set about establishing his orchards, a rival human system for propagating fruit trees was already well established. No one knows who first discovered that it was possible to graft trees together by carefully tying a branch or cutting from one growing tree on to the roots or stem of another. The technique was definitely used in ancient Greece and Rome, where apples, transported from Asia via the silk route, were widely propagated as a favourite fruit. In 323 BC Greek writer Theophrastus, a pupil of Aristotle and a keen botanist, wrote a comprehensive practical treatise that included advice on the best techniques for grafting trees.

Trees propagated in this way avoid the lottery of sexual reproduction's genetic mix 'n' match. Choose a specimen that produces the shape, form and flavour of fruit you like, then graft one of its branches on to the roots of another tree of the same (or similar) species and, hey presto! Identical fruit as produced by the first tree will grow back every time. This propagation technique works for most fruits and flowers that grow on vines or trees, from peaches and almonds to apples and pears (see also roses, page 354, and grapes, page 307).

Despite this, Chapman insisted on growing his trees from seeds that originated in Europe. The upside of Chapman's intransigence was that those seeds that survived were those best suited to the conditions of their new American home. The downside was that it was impossible to predict the quality, quantity, size, shape or form of fruit from one tree to the next. Some were small and sour, others were large and sweet. Some were susceptible to disease, others mostly immune. There was absolutely no standard or consistency – just as nature intended.

For more than a century these quirks didn't matter because American settlers never ate Chapman's apples. Instead, they were crushed, fermented and turned into a nourishing refreshing brew called cider – a beverage consumed even in preference to water (as was beer in early modern England, see page 284) because it was a clean, healthy, safe drink for all the family. But when at the start of the twentieth century apples began to be consumed as food, the fate of the apple tree changed dramatically. A marketing campaign was launched to get Americans to '*eat an apple a day – it keeps the doctor away*' and the onset of Prohibition, which made the consumption of alcohol illegal, stunted the appeal of cider. Since human consumers judge the fitness of food on its visual symmetry, colour and proportion random sizes, shapes and variable tastes were no longer acceptable. Grafted trees that produce predictable fruit became all the rage. They have remained so ever since.

As a result, the number of varieties of apples grown across America (and now the whole world) has declined dramatically – in defiance of Chapman's best-practice plantation programme. Out of the 7,500 known apple cultivars, almost all of the fifty-five million tonnes of apples grown worldwide each year are derived from a core stock of just six varieties, specially selected for their looks, shape and sweetness.

The risks of farming so few mass-produced varieties (see also water mould, page 40) resulted in a bacterial near-miss in the late 1980s when a disease called fire-blight (caused by the bacterium *Erwinia amylovora*) ripped through the fruit orchards of North America. As a result, many growers faced collapse in a similar way to French vintners just over a century before (see pages 309). Only thanks to a revolutionary approach to pest control was resistance to the disease established. Researchers at Cornell University successfully pioneered the use of a 'gene gun' to transfer an antibacterial gene from a species of giant silk moth (see page 266) to an apple tree. Cuttings from this tree have since allowed farmers to propagate fire-blight clones in a bid to immunize their orchards.

The need for such techniques would have American hero Johnny Appleseed spinning in his grave. Apples earn their place as the third most popularly grown fruit (after bananas, see page 390, and oranges, see page 386) but artificial genetic intervention has become an ingredient vital to ensure the survival of inbred crops specifically designed to satisfy the innate human desire for beautiful healthy-looking food.

Vanilla

> FAMILY: ORCHIDACEAE
> SPECIES: VANILLA PLANIFOLIA
> RANK: 97

How the dried beans of a Central American orchid became one of the most popular flavourings in the modern world

ANNE SHIRLEY, AN eleven-year-old red-haired orphan, was utterly determined not to miss her Sunday school picnic tea-party. But she had a problem. Her guardian Marilla was under the distinct impression that Anne had stolen a precious amethyst brooch. As punishment Anne had been banished to her bedroom until she owned up to the crime. So desperate was precocious little Anne not to miss the party that she made up a false confession: '*Punish me any way you like,*' pleaded Anne to Marilla, '*but please, please let me go to the picnic! Think of the ice cream . . . For anything you know, I may never have the chance to taste ice cream again.*' Later that day Marilla found her brooch, which she had carelessly mislaid; the confession was exposed as false and Anne skipped off gleefully to the picnic after all. '*Words fail me to describe that ice cream,*' Anne told Marilla after she returned home to Green Gables, '*I assure you it was sublime.*' (From *Anne of Green Gables*, 1908, by Lucy Maud Montgomery.)

The popularity of this cool, refreshing and typically irresistible dessert can be put down to a powerful flavouring extracted from the cured beans of an exotic species of orchid called vanilla. Native to Mexico, the plant was known only to the Totonac and later Aztec people of the region until Spanish conquistadors led by Hernán Cortés arrived in the early sixteenth century. Mixed with hot chocolate, vanilla sweetened the flavour of this drink for emperors and the nobility. Taxes in the form of vanilla beans were paid as tributes to the growing Aztec empire in the fifteenth century, but only those who lived in the surrounding hills knew the complex art of cultivating the plant and extracting its precious essence.

European attempts at growing the plant in other parts of the world ended in abject failure for as long as 300 years. Even when the plant flowered, for some reason it never seemed to produce its delicious-smelling pods anywhere other than in its native Mexican habitat. Until the mid-nineteenth century supplies were therefore restricted to a small area where the orchid grew naturally, and the European fancy for adding vanilla to cocoa (an addition generally preferred to the Native American alternative, chilli, see page 280) was strictly limited to an aristocratic elite.

The vanilla orchid's mysterious lifestyle was finally exposed thanks to two discoveries that followed each other in quick succession. The first came in 1836, when Belgian botanist Charles Morren (1807–58) spotted a particular species of black bee buzzing around a vanilla orchid that decorated the patio of the house where he was staying in Veracruz, Mexico. After close observation he saw that in the process of finding nectar these pollinators, called Melipone bees, lifted up a transparent flap before depositing their load of pollen. This flap, called a rostellum, is nature's way of ensuring that only the pollen from other orchids visited by these bees gets rubbed on to the style, thus avoiding the hazards of self-pollination and inbreeding. Vanilla orchids produced their aromatic beans only in Mexico because this was the one place inhabited by Melipone bees. Flowers cultivated elsewhere remained unfertilized. They simply withered and died without ever bearing fruit.

Perhaps of greater surprise to vanilla ice cream lovers is that they enjoy their favourite flavour only thanks to the ingenuity of an African slave called Edmond Albius (1829–80). At the tender age

of twelve this orphan boy developed a quick ingenious way of pollinating vanilla flowers by hand, without the need for Melipone bees.
Using nothing more than a thin bamboo cane and a quick flick of
the thumb, Albius found he could manually pollinate up to 1,000
orchid flowers an hour. Vanilla orchid production suddenly became
big business. So successful was Albius's pollination technique that
it is still used today. The Indian Ocean islands of Réunion (where
Albius lived) and Madagascar, as well as South America and the West
Indies, became new habitats in which vanilla orchids were subsequently cultivated and harvested. By 1898 more than 200 tonnes
of cured beans were being exported from Madagascar alone, still the
world's largest vanilla producer.

Given the enormous demand for vanilla flavouring today, not
just in chocolate bars and ice cream, but also for use in perfumes
and aromatherapy oils, it might be expected that the cultivation
of vanilla orchids would have followed the same pattern of intensive agriculture as other high-demand crops such as sugarcane (see
page 206), roses (see page 354) and coffee (see page 283). But no.
The lasting impact of the genes of this awkward-to-cultivate orchid
manifested itself in a spur to human scientific endeavour rather than
in the reality of a reshaped landscape. The flavour of vanilla proved
so popular in Europe and America that only thirty-three years after
Albius discovered how to propagate the plant, German scientists
worked out how to recreate the chemical structure of its flavour-
producing compound, vanillin, artificially. Of the 12,000 tonnes of
flavouring used in vanilla-flavoured products today, only about 10
per cent comes from genuine, naturally cured vanilla beans. The rest
is produced artificially using a base resin derived from the leftovers of
woodpulp processing or from a chemical called guaiacol, extracted
from the tar of beechwood.

Sometimes the economics of cultivation are such that it is cheaper
for humans to turn their hands to organic chemistry rather than
agriculture (see also synthetic rubber, page 265) to satisfy popular

demand. Nevertheless, had humans not experienced the taste and smell of natural vanilla, they would never have tried, let alone been able to copy, nature's chemistry. While the widespread use of artificial flavourings may be worse for the dissemination of the genes of *Vanilla planifolia*, it is probably a lot better for the health of the global ecosystem, dominated as it is by so many people with a love of sublime ice cream.

Lavender

FAMILY: LAMIACEAE
SPECIES: LAVANDULA ANGUSTIFOLIA
RANK: 93

*An alternative-therapy-cum-placebo that has wafted
its way into the history books*

THAT HUMANS HAVE a lousy sense of smell was not lost on
Charles Darwin, who said in his book *The Descent of Man* (1871):

> '*The sense of smell is of the highest importance to the greater number
> of mammals – to the ruminants in warning them of danger, to the
> carnivore in finding their prey . . . but the sense of smell is of extremely
> slight service, if any, to men . . . He inherits the power in an enfeebled
> and so far rudimentary condition, from some early progenitor . . .*'

Therefore only plant species with the very strongest odours were ever
likely to grab people's attention through their olfactory pipe – the
nose. Pretty flowers that grow around the Mediterranean are two-a-
penny, but what really distinguishes the herb purple lavender from
most other plants is its unforgettable fragrance.

People living in Europe and North Africa from the days of ancient
Egypt to the present have shared a common belief that something
as aromatic as lavender must have a therapeutic, medicinal or other
powerful effect. The catalogue of uses this single herb has been put
to, and the many effects ascribed to it, are extraordinarily diverse.
Traces of lavender fragrance were found by archaeologists when
they rummaged through the treasures of Tutankhamen's tomb after

its discovery by Howard Carter in 1922. Lavender oil was used in Egyptian and Phoenician purification ceremonies and was added as an ointment during the all-important process of mummification. Cleopatra, the reputedly beautiful last pharaoh of Egypt, is said to have seduced Roman generals Julius Caesar and Mark Antony with the aid of perfumes laced with lavender.

The name 'lavender' derives from the regular use of this strong fresh-scented herb by the people of ancient Rome. They added it as a fragrance when washing their linen or taking a bath – *lavare* is a Latin word meaning 'to wash'. According to the Gospel of St Luke, Mary Magdalene, a disciple of Jesus Christ, anointed his feet with oil that contained spikenard, a herb thought by some scholars to have been a type of lavender. As with so many other products of the natural world (from grapes, see page 307, to rabbits, see page 344) the Romans transported powerful-smelling lavender all over their empire, where it took root in Spain, Britain and France. By medieval times it was being used as a miracle cure for a wide range of ailments and diseases from head lice to warding off the devil. Author, linguist, scientist and philosopher Hildegard of Bingen (1098–1179), who founded a monastery on the river Rhine, was under no illusions as to the mystical power of the herb: '*If a person with many lice frequently smell lavender, the lice will die . . . It cures very many evil things and because of it malign spirits are terrified.*' The use of lavender-oil-soaked blotting paper against head lice continued as a popular remedy until as recently as the 1870s.

Fragrant lavender became associated with royalty from the time of Queen Elizabeth I of England (ruled 1558–1603), who believed its medicinal properties helped cure her frequent head-aches. Henrietta Maria, wife of King Charles I, introduced lavender soaps and potpourri from France just as the French perfume industry was establishing itself under the patronage of Catherine de Medici (1519–89), mother of a trio of French Kings: Francis II, Charles IX and Henry III. Renato Blanco, an Italian perfume expert and personal employee of Catherine, is credited as being the first professional perfume maker in modern times. His laboratory was connected to Catherine's palace via a secret passageway to prevent others from knowing the exact compositions of her sweet-smelling fragrances. Soon after this, the Pilgrim Fathers took lavender with them to the Americas, where it thrived despite the harsh winter weather. By the time of Queen Victoria (ruled 1837–1901) lavender had become a royal obsession. It was used in everything from eau-de-toilette to a disinfectant solution for washing down the floors and furniture of her many royal palaces. She even appointed Lady Sarah Sprules '*Purvey of Lavender Essence to the Queen*'.

But the powerful scent of lavender wasn't just the focus of people's desire to smell good. It was also thought that something so fragrant could fight off plague, disease and infection – a popular belief that originated in ancient Rome but which was revived during the plagues that struck Europe from the fifteenth century. In London during the Great Plague (1665–6) people tied bunches of lavender around their wrists, hoping the smell would ward off infection. The use of lavender oil in solution as an antiseptic was widespread even into World War I before the advent of drugs such as penicillin (see page 326).

In modern times lavender not only remains a major constituent of perfume (see plate 11, top) but has also become the premier essential oil in the global market for alternative medicine. Aromatherapy – a practice that emerged in the first half of the twentieth century – claims to be able to improve mood, health and spiritual wellbeing. It was developed by René-Maurice Gattefossé (1881–1950), a French chemist who specialized in manufacturing perfumes. While working in his laboratory Gattefossé accidentally burnt his hand and, to ease the pain, instinctively thrust it into the nearest vat of liquid. The container was filled with lavender oil. The remarkable speed with which Gattefossé's hand healed inspired him to explore more fully the restorative effects of naturally occurring aromatic compounds.

His book *Aromathérapie* (1937) became the cornerstone of the modern alternative medicine movement, which is now a multi-billion-pound business.

Modern Western science has little if anything to say about the medicinal efficacy of lavender apart from its obvious properties as a highly pungent fragrance. Yet it is beyond dispute that for thousands of years this herb has convinced people of all social classes that its strong odour has beneficial medicinal or aphrodisiac effects. Placebo or ancient remedy, lavender's ability to power its way through the mostly dulled capacities of the human olfactory system undeniably accounts for its worldwide success.

Black Pepper

> FAMILY: PIPERACEAE
> SPECIES: PIPERACEAE NIGRUM
> RANK: 91

A vine that people were convinced was essential for their personal wellbeing

BLACK PEPPER, a vine that grows up the sides of trees, is native to the monsoon region of south-west India (Kerala). Its fruit grows in clusters of red berries which, once harvested, are left to shrivel in the sun, turning them black. When crushed, these light, easy-to-transport peppercorns can be added to any sort of food, giving it a hot, exotic, spicy flavour.

The triumph of mind over matter best accounts for the historical popularity of black pepper. Ancient Greek philosophy was heavily influenced by the Aristotelian concept that the secret to maintaining good human health lay in the proper intermixing of the four essential elements that comprised the natural world – earth, air, fire and water. Scholars such as Hippocrates (460–370 BC) and Galen (AD 129–200) said that when applied to the human body these elements manifested themselves as four humours: black bile (melancholy), yellow bile (choleric), phlegm (phlegmatic) and blood (sanguine). Until the birth of modern Western science in the late seventeenth and early eighteenth centuries, medical practice was primarily concerned with finding the correct balance between these humours to aid wellbeing.

Pepper, it was widely believed, was vital to this process. It was one of the few natural ingredients with a hot dry effect that could temper a diet of sanguine (red) meat. Today's rump steak with peppercorn sauce is a modern relic of a medieval passion for spicy food, long since forgotten with the emergence of a taste for sugar (see page 206), which became widely available from the seventeenth century. The need to balance the effects of different foods on the stomach was thought vital for healthy living. That's why spices like pepper were in such demand by those who could afford them, from monks to merchants, and kings to courtiers. *Circa Instans*, a medieval recipe book written *c.* 1160, lists 270 substances thought to help balance the four essential humours. The spiciness of black pepper was especially effective, it said, for removing wet phlegm from the chest, as a remedy for asthma. Pepper would also heal bloody sores when ground into a powder.

Such humours didn't regulate only physical health. Sexual appetite was stimulated by the use of spices, according to an ancient Greek work translated from Arabic into Latin by eleventh-century Benedictine monk Constantine the African. According to the text *De Coitu* ('On Sexual Intercourse'), a man's impotence was due to an excess of liquid humours (phlegmatic and melancholic), which could be corrected by the use of hot spices such as pepper, ginger and cinnamon.

The fact that spices such as pepper came from a mostly unheard-of corner of the world only added to their mystique. Scarcity and myth – possibly perpetrated by enthusiastic Islamic merchants – were additional psychological factors that helped make this flavour seem desirable to medieval palates. Isidore of Seville (AD 560–636), Archbishop of Milan, was the author of another authoritative natural encyclopaedia, called the *Etymologiae*. In it he explained how Indian pepper trees were 'guarded' by poisonous snakes. He said that in order to harvest the valuable berries, natives would light fires between the trees to ward off the scary serpents. His colourful explanation also explained why peppercorns were always black and shrivelled.

European determination to gain direct access to supplies of pepper was boosted by the thirteenth-century globetrotter Marco Polo's popular account of the bountiful supplies of pepper produced in the spice island of Java (see plate 14, bottom). He described it as a rich island: '*producing pepper . . . and all the precious spices that can be found in the world. This is the source of the most spice that comes on to the world's markets. The quantity of treasure . . . is beyond all computation.*' Explorers such as Christopher Columbus and Vasco da Gama were ready to risk everything to find a route to these fabled lands. When da Gama finally reached Calicut on the west coast of India in May 1498, he began negotiations to secure spices even though he had little of value to trade in return (see also page 379). He intended his overtures to be taken seriously. On his return three years later he devastated the Indian town with his cannon, imprisoned the locals and burned them alive before returning to Portugal with more than one million kilos of pepper, equivalent to what Venice obtained during an average year. By 1530 the Portuguese had established a capital at Goa with a string of fifty forts along the west coast of India and a fleet of more than one hundred ships. The era of unequal trading between the nations of Europe and the East had begun.

As a result of these adventures, modern historians are minded to credit pepper with some of the most extraordinary achievements in human history. According to one: '*The starting point for the European expansion out of the Mediterranean and the Atlantic continental shelf had nothing to do with, say, religion or the rise of capitalism – but it had a great deal to do with pepper.*' Another cites spices, of which pepper was by far the most popular, as having '*dramatically affected history because they launched Europe on the path to eventual overseas conquest whose success and failure affects every aspect of contemporary world politics*'.

In the last 500 years black pepper, a catalyst for European conquest, has retreated into relative obscurity. During the seventeenth century the rise of new medical knowledge and the arrival of other more addictive substances in Europe (such as laudanum, see page 322; sugar, see page 206; tea, see page 287; and coffee, see page 283) reduced the demand and allure of pepper from a medicinal necessity to a more occasional culinary flourish. But by then its impact had been felt worldwide. From the discovery of the Americas to the colonization of east Asia, world history, as a result of pepper, tacked into a new headwind.

Lotus

> FAMILY: NELUMBONACEAE
> SPECIES: NELUMBO NUCIFERA
> RANK: 99

A flower that has inspired the search for inner peace and is now regarded as a model material by Western science

FOR 3,000 YEARS people from the Far East have held up the sacred lotus flower as the ultimate model of natural perfection. Hinduism, the most ancient existing human religion, and Buddhism, its more recent offshoot, draw great inspiration and power from the shape, form and function of this flower (see plate 15) – not to be confused with the Egyptian blue lotus, which belongs to a different species of water lily.

On one level, the lotus flower, like most other flowers, represents visual, symmetrical perfection – its unfolding petals spreading out evenly to present the plant as a genetically fit and attractive prospect for birds, insects or other passers-by. But it is the habitat of the lotus, a perfectly clean flower that arises as if by magic from murky muddy water, that has most captivated the human imagination. In the Bhagavadgita, the holiest of Hindu scriptures, the lotus flower is used to demonstrate what humans should strive for in order to achieve inner perfection: *'One who performs his duty without attachment, surrendering the results unto the Supreme Lord, is unaffected by sinful action, as the lotus is untouched by water.'*

Shortly after his enlightenment, Siddhartha Gautama, known as the Buddha (*c.* 563–483 BC), showed a lotus flower to the audience that had come to hear him preach. All he did was lift the flower towards the assembled crowd – that was it. The Flower Sermon, as the occasion has been remembered, contained no spoken words. Part way through a man in the assembly smiled at the Buddha, showing that he understood the potency of the message being conveyed. Mahakasyapa went on to become one of the Buddha's most revered disciples, and the Flower Sermon, as a way of reaching inner perfection that transcends words, is now considered the birth of the branch of Buddhism called Zen. The noble path towards human perfection in Eastern culture concerns stilling the mind, freeing it from desire, angst, worry or guilt. Just looking at the lotus flower, suggested the Buddha, without speaking a word, was key to finding this inner calm. The approach was in stark contrast to the Greek philosophical tradition, founded by Socrates (470–399 BC) and his pupil Plato (428–347 BC), which sought perfection in human society by stimulating argument, rhetoric and debate. It is in this culture that Western society has its roots.

Tibetan monks continue the Buddha's tradition today through the symmetry of a mandala – a concentric image – which they use as an aid to meditation and trance. These geometric images are created out of multicoloured sand, funnelled into lines and shapes using tubes and scrapers. At the centre lies the perfect image of a lotus flower. Ancient Hindu traditions for maintaining human spiritual health are focused around the seven chakras, lotus-shaped vortices that are thought to channel and circulate *prana*, the universal life-giving force, in and around the body.

Nature's mirror, celebrated in Eastern cultures for thousands of years in the form of a lotus flower, has recently begun to be appreciated by Western scientists in their own pursuit of perfection based on the material world. Biomimetics is a relatively new branch of science that aims to copy the way nature solves problems and apply it to the human world in a sustainable way.

In the early 1970s German botanist Wilhelm Barthlott uncovered the mystery of how lotus flowers are able to stay so pristine despite growing in some of the muddiest waterlogged habitats on Earth. After careful observation of the plant's leaves through an electron microscope he discovered they were covered in tiny protrusions called papillae. This design ensures that when a droplet of water falls on to a leaf, its surface tension stays intact. This causes the droplet to roll across the surface of the leaf, gathering dirt or other unwanted particles along the way, before falling off the plant altogether. Even sticky honey and glue are unable to attach to the leaves of the hydrophobic lotus plant. This simple self-cleaning system protects the plant against attack from bacteria, fungal spores and other harmful pathogens.

The science of biomimetics uses the study of natural systems to transform the world of artificial materials science. Dirt-repellent self-cleaning coatings embedded with rough microscopic surfaces similar to those found on lotus leaves are being trialled on products including surface paints, roof tiles and waterproof clothes and shoes.

Today the sacred lotus can be appreciated in all its glory by both Eastern and Western cultures. Whether seeking inner peace or material perfection, many millions of people continue to find the lotus a source of natural inspiration.

Chilli Pepper

> **FAMILY: SOLANACEAE**
> **SPECIES: CAPSICUM ANNUUM**
> **RANK: 92**

A brightly coloured, easy-to-grow, adaptable
fruit with a fiery flavour

WHEN A WARNING light suddenly illuminates in an airline pilot's cockpit does it mean there is a genuine malfunction with the plane's systems or is it simply a faulty sensor? A similar question might occur to a person who accidentally bites into a raw chilli. Is the searing pain that explodes in the mouth causing actual bodily harm or is this fruit somehow triggering a powerful false alarm?

Chilli peppers are thought by archaeologists to be among the first crops to have been domesticated by humans living in the Americas. Capsicum grains, used in cooking, have recently been found in south-west Ecuador dating back to 4000 BC. Their discovery, on the opposite side of the Andes mountain range from their natural habitat in Peru, suggests that chillies were a highly valued and widely traded supplement for early farming people who enjoyed eating these fruits to spice up their lives.

Quite why humans enjoy the flavour of chilli peppers is something of a mystery. These plants evolved to appeal to the palates of birds, not mammals. The active chemical produced in the fruit of the chilli plant, capsaicin, evolved as a defence against fungi and grazing herbivorous mammals. Rather than stimulating one or more of the five standard tastes, this compound binds to the same receptors in a mammal's mouth as those that respond to hot temperatures or tissue damage, triggering a sensation of heat and pain that persuades a mammal's brain that its mouth is being ripped apart by a surge of something violently hot to the touch. As anyone with a fancy for vindaloo curry knows, the result is an increase in heart rate and sweat. This feeling is a complete illusion, however – a natural false alarm. Devour a hot chilli and the temperature inside your mouth will be no different from how it was before you took the first bite. Birds are insensitive to the chemical, allowing them to eat the seeds of raw chillies without discomfort. Yet mammals recoil with pain.

There is a simple evolutionary explanation. When chilli seeds are eaten by herbivores they get damaged by their grinding molars. But when the seeds are eaten by birds they pass through the digestive tract unharmed, allowing them to be spread successfully in their droppings. Therefore, over millions of years, natural selection has seen to it that the bright yellow and red colours of mature peppers attract birds, while the hot, choking chemical capsaicin wards off grazing mammals. So why did early American agricultural humans begin to cultivate these obnoxious fruits as a food of *choice*?

One reason may involve the side-effects of capsaicin. In humans this substance triggers the release of endorphins – chemicals that reduce stress, relieve depression and raise the spirits. Also humans are unlike other mammals in that they have learned how to cook. Sizzling peppers alongside other foods mixes the hot taste of chilli, both diluting and enhancing its effects. A third factor may be psychological. It is a well-known phenomenon that humans enjoy the sensation of pushing their bodies beyond standard operating limits, as when thrill-seeking by riding a roller coaster (see also cannabis, page 292). Experiencing pain without actually being physically harmed is thought by some experts to have similar appeal.

It wasn't until the arrival of European explorers with the voyages of Christopher Columbus in the late fifteenth century that chillies had the opportunity to experience habitats beyond their native

American origins. Spanish and Portuguese people became their new-found champions, calling them chilli peppers after the similarly spicy pepper-like effect they had on the palate. All the medicinal benefits afforded to black pepper (see page 370) as a way of balancing the body's humours were now also bestowed on chillies, as a Spanish physician writing in the 1570s was eager to explain: '*It doth comfort much, it doth dissolve the winds, it is good for the breast and for them that be cold of complexion it doth heal and comfort, strengthening the principal members.*' Spanish people began to use chilli peppers in their cooking instead of the traditionally expensive and difficult-to-obtain black pepper. Meanwhile, Portuguese explorers settled on the west coast of India soon after the pioneering voyages of Vasco da Gama (see pages 372). By the 1530s they had established India's first chilli pepper plantations. Seeds were probably brought from Brazil via Lisbon to India. Once cultivated in the traditional spicy-black-pepper-growing region of western India, chillies became an instant hit – a perfect antidote to the blandness of rice.

Indian spice farmers dried and crushed these new culinary delights, turning chillies into easy-to-transport powder. By the late sixteenth century capsicums grown on the west coast of India were being traded along the medieval spice routes into Persia and around the Black Sea, where they were quickly incorporated into Middle Eastern and Turkish cuisine. From there it was just a short hop over to central Europe, completing the spice's global journey. Following the Turkish invasion of Hungary in 1529, ground chilli powder in the form of paprika became a new European favourite, giving additional colour, taste and flavour to everything from pepperoni sausages to Hungarian goulash soup. Today's European and American consumers who dislike the spiciness of traditional chilli peppers can enjoy the common bell-shaped pepper, *Capsicum annuum*, which has been bred to contain little or no pain-producing capsaicin.

With as many as thirty varieties catering for all manner of tastes, from benign to fiery hot, the global adoption of chilli peppers into so many different cuisines will ensure these fruits continue to be widely propagated for as long as humanity remains a dominant global evolutionary force. It's a somewhat ironic result for a fruit that succeeded in the wild by developing defences against being eaten by mammals with molars.

Grass

> **FAMILY: POACEAE**
> **SPECIES: POA PRATENSIS • LOLIUM PERENNE •**
> **AGROSTIS STOLONIFERA • FESTUCA RUBRA**
> **RANK: 24**

*A velvet carpet of green that humans have cultivated to
please the eye and soften the fall*

MODERN HUMANITY DEPENDS on grasses for its survival.
These vigorous plants contain the third largest number of species
among the flowering plants (angiosperms) after orchids and asterids
(the daisy family). Several have already featured in our top one
hundred (see wheat, page 202; rice, page 236; sugarcane, page
206; maize, page 241; and bamboo, page 153). There are, however,
other types of grass that have been even more widely cultivated, not
because they are good to eat or useful for construction, but because
they appeal so greatly to our senses of sight and touch.

Gardens, lawns, city parks, football pitches and golf courses are
just a few of the countless horticultural features dominated by
cultivated species of cut green grass. According to recent research
conducted by NASA scientist Cristina Milesi, satellite images reveal
that there are *three times more* acres of grass lawns in the United
States than irrigated corn (maize). The total surface area covered by
cut grass (including golf courses and sports fields) is estimated to
cover an area in excess of 128,000 square kilometres.

Gardening is as old as recorded history. Ancient Chinese records
tell how it was in a garden that Leizu reportedly made her silkworm
discovery (see page 266) and it was in a garden that God spoke

381

to Adam and Eve, warning them not to eat of the forbidden fruit, according to Jewish scribes who wrote the Torah in about 500 BC. But the roots of cultivating a garden dominated by a velvety grass lawn go back to an aristocratic fashion that caught on in eighteenth-century England. Lancelot 'Capability' Brown (1716–83) designed more than 170 gardens, many of which surrounded imperial Britain's finest country houses. Brown's passion was to use vast swathes of closely cut undulating grass to provide a soft, pleasing velvet-like carpet that drew an onlooker's eye towards the central artificial architectural masterpiece. Newly sown green grasses became a symbol of aristocratic status, wealth and prestige because keeping such extensive areas of closely cut lawns required a veritable army of garden staff equipped with an arsenal of specialist cutting tools including lifters, edgers and scythes (see plate 14, top).

As European social ideals were transplanted to the newly independent United States of America, the finely manicured lawn became that new nation's most potent symbol of man's triumph over nature. President George Washington (in office 1789–97) cherished the lawn at his Mount Vernon estate, which symbolized his country's ongoing conquest of a hostile landscape into a cultivated continent civilized by man. In his diary, dated 1785, he wrote of his pride for his grass: '*I sowed what was levelled and smoothed of it with English grass seeds; and as soon as the top was so dry as not to stick to the roller, I rolled and cross-rolled it.*'

Washington's desire to see his lawn criss-crossed and rolled with stripes underlies the power of grass to please people's love of visual geometrical perspective and symmetry in nature. What better display of the genetic fitness to govern than a meticulously groomed well-rolled lawn? It was the perfect adornment for a ruler intent on stamping his authority on nature and society. The extensive well-cut grasses that still surround the American presidential residence, the White House, were themselves inspired by Washington's Mount Vernon estate.

On Beauty

American garden designer Andrew Jackson Downing (1815–52) took Washington's love of finely cut freshly rolled grass to a mass audience. In his many writings about the theory and practice of landscape gardening Downing made specific recommendations about the types of grass best suited to make the perfect lawn which, he said, should be cut with an English scythe edged like a razor: '*No expenditure in ornamental gardening,*' he wrote, '*is productive of so much beauty as that incurred in producing a well-kept lawn. It is a universal passport to admiration.*' Popular admiration for perfectly cut grass exploded soon after Jackson's death with the invention of the mechanical lawnmower. American inventor Elwood McGuire, from Indiana, pioneered the first human-powered machines in 1870. Now anyone could show off their own ability to civilize nature with a weed-free, finely cut, perfectly striped lawn surrounding their ideal home on all four sides.

Golf, cricket, lawn tennis, croquet, football and rugby were all sports that took off towards the end of the nineteenth century thanks chiefly to the powerful combination of mechanical cutting equipment and new varieties of soft spongy grass. Such surfaces not only looked good but also proved perfect cushioning for everything from the bounce of a golf ball to a fall from a badly timed tackle. In fact, so ideal was cut grass as a surface for European and American sports that by the early twentieth century agronomy – the business of cultivating, breeding, growing and protecting grasses – became established as its own branch of science, leading to the founding of the American Society of Agronomy in Chicago in 1907. Twenty-one years later an otherwise unlikely alliance between the US Department of Agriculture, the US Golf Association and the Gardening Club of America pioneered the growing of as many as 500 separate plots of different varieties of grass in an effort to find a combination that consistently produced the best terrestrial coverings for human deployment and enjoyment.

Such research confirmed that the attempt at breeding a perfect species of grass was a false pursuit. Much better was to sow areas with a mixture of varieties, some of which are tolerant to drought, others tough enough to withstand being smashed by balls or churned up by feet. As a result, a mixture of species from four families of grasses has most commonly been used since the early twentieth century to create the purpose-sown lawns, parks, playing fields, golf courses and grasslands of the modern world. Bluegrasses (such as *Poa pratensis*) are well suited to shady, colder temperatures while rye grasses (such as *Lolium perenne*) work best in dry conditions. Others such as *Agrostis stolonifera* and *Festuca rubra* provide a fine, soft texture, ideal for bowling lawns and putting greens.

Between 1947 and 1964 an average of more than 1.25 million new homes were built in the United States every year, each one established within its own garden plot. Social convention dictated that they were surrounded by cut grass. From 1963 the air-cushioned Flymo, invented by Swedish engineer Karl Dahlman, allowed even females to cut grass easily because it was '*a lot less bovver with a hover*'. Meanwhile, a huge industry sprung up around the ideal American suburban home. Automatic sprinklers, grass seed, fertilizers, pesticides and machinery for tending to the upkeep of grass all aided the growth of a new multi-billion-dollar industry. In Britain the post-war building boom was no less spectacular – with an estimated total of sixteen million lawns being rolled into regular service by 1989.

Brown, dormant grass looks unattractive so humans tend to insist on extensive watering to keep their lawns looking healthy and green. NASA's Milesi has calculated that as much as 900 litres of fresh drinking-quality water are needed per person per day to keep America's lawns looking fresh. This is a significant factor, she says, behind the rapidly falling water table along North America's eastern Atlantic seaboard.

Without irrigation or artificial fertilizers most of America's cut grass would perish. The only places lawn-quality grass could grow in the US without human assistance are a few areas in the north-east and the Great Plains. Just picture how different a world without cut grass would look. Would mankind's relationship with nature be different today if deprived of its most conspicuous stamp of human authority? Would the mass adoption of competitive sports from golf to football have been possible without this supple, robust, naturally rejuvenating surface? Could people bear to live in urban environments if totally deprived of their parkland patches of velvet green?

Orange

> **FAMILY: RUTACEAE**
> **SPECIES: CITRUS SINENSIS**
> **RANK: 95**

*A globally dominant, brightly coloured berry that helps
compensate for an ancient genetic weakness*

EBOT MITCHELL, a housewife from Cornwall, England, wrote a remarkable book in the early eighteenth century which recently surfaced in the attic of a Gloucestershire home. It detailed more than one hundred household recipes, one of which was found to be the first known description of a cure for a disease that has plagued human beings from at least the time of ancient Greece.

Scurvy is a particularly nasty illness, traditionally contracted by sailors on long voyages. Symptoms include excessive bleeding, loss of teeth and hair, bone deformation, hallucinations and blindness, eventually leading to death. Mitchell's book described an effective remedy nearly fifty years before James Lind, the Scottish naval officer usually credited with discovering a cure. Her answer, like Lind's, was perfectly simple: always eat plenty of citrus fruits, like oranges. This is because they are rich in vitamin C.

Vitamins occur naturally in a wide range of fruit and vegetables, but vitamin C is highly concentrated in the small family of fruits called citrus. By far the most popular species of citrus fruits today are oranges. Sweet varieties called *Citrus sinensis* are grown in huge orchards all over the world from Florida to Spain, California to China and Australia to the South African Cape. Other citrus crops of significance, but far less intensively grown, include other varieties of orange (in particular mandarins) but also lemons, limes and grapefruits.

It is something of an evolutionary quirk that humans need to eat a diet of fresh fruit stuffed with vitamin C at all. Most animals have enzymes that automatically synthesize vitamin C to keep their bodies healthy – in the same way humans automatically manufacture their own vitamin D in their skin using little more than sunlight. But for some ancient evolutionary reason primates (including humans) have lost this capability when it comes to the manufacture of vitamin C. Instead, it is vital for their good health to find naturally occurring external sources to supplement their diets.

Not that this was a problem for most of human history. As hunter-gatherers, eating wild berries ensured vitamin C deprivation never occurred in the Stone Age. Only with the emergence of settled human societies did scurvy begin to become a problem, and then only if people failed to eat sufficient quantities of fruit and veg. The disease became most acute from the sixteenth century onwards when European maritime explorers went on epic voyages all over the world (see page 388). Not realising their diets had to be supplemented with sources of vitamin C, many sailors died appalling deaths as a result of a mysterious disease called scurvy.

Oranges, Ebot Mitchell's solution to the scurvy problem, have been enjoyed by humans for almost all of recorded history itself. Their first recorded appearance was in China, hence their botanical name *Citrus sinensis*. The sweet oranges commonly grown throughout the world today are a hybrid cross between two species, the grapefruit-sized *Citrus maxima* and the mandarin *Citrus reticulate*.

It is still unclear exactly which people transported oranges across Asia to Europe and when. The Jewish Torah makes no obvious references to citrus fruits, although figs, olives and dates appear throughout. But it is clear that by the time of the later Roman empire species of orange were being grown around the Mediterranean. They were regarded mostly as exotic, luxurious and usually ornamental fruits because these varieties had a much more bitter taste than the sweet oranges popularly grown today. The same sour varieties were also spread by Arabs following the Islamic conquests of the eighth century. They were used by herbalists such as Avicenna (AD 980–1037) to make a type of medicinal syrup called *alkadare*. The fruit therefore spread throughout Egypt and North Africa to Spain, Sardinia and Sicily. It was only when sixteenth-century Portuguese maritime explorers brought back seeds from the sweet orange hybrids grown in China (*C. sinensis*) that the fruit began to feature more fully in the minds and stomachs of people living in the Middle East, Europe and North Africa. By 1646 sweet orange trees were growing in Italy, introduced from Portugal (but originally imported from China).

Christopher Columbus and other European colonists took seeds of the sweet orange with them in their overseas survival kits. They were planted on the island of Hispaniola (Haiti), where Columbus founded his first colony in 1493. Spanish settlers then took the fruit to Florida some time between 1513 and 1565, when their first colonies were founded around St Augustine. Once introduced, these fruits relished the subtropical American climate. Wild orange and other citrus fruit groves also sprang up around Florida's many lakes.

Meanwhile, the delightful taste of sweet oranges grown around the Mediterranean started to make its way into preserves for transportation and consumption further north. Marmalade became a favourite at British and French breakfast tables from the eighteenth century. A dedicated factory opened on the east coast of Scotland in 1799 that specialized in turning the whole fruit of sweet Seville oranges into delicious Dundee marmalade. At about the same time greenhouses called orangeries began to grace aristocratic estates as *C. sinensis* became fashionable to cultivate in cold climates because of its attractive golden fruits.

The twist that turned sweet oranges from an exotic aristocratic fancy into the mass-consumer product we know today came about thanks to the nineteenth-century trilogy of education, transportation and refrigeration. Knowledge of the health benefits of eating citrus fruit that started with the cure for scurvy, as written about by Mitchell and Lind, was quickly followed by the railway revolution that gathered steam in the mid-nineteenth century. By 1869 freshly picked oranges grown in sunny climates such as California (planted as a result of the gold rush 1848–55) could be transported quickly by train all the way to New York. Within twenty years the trains were upgraded with refrigeration, preserving the fruits' freshness. By 1892 the first steamers loaded with oranges set off for London. Queen Victoria, when she tasted one, is said to have pronounced it '*palatable*'.

Advanced mechanical processing techniques emerged soon after World War I that automatically converted fresh oranges into easy-to-transport concentrated juice that could be reconstituted close to its place of consumption. Such innovations mean that as human populations have surged over the last sixty years people's genetic weaknesses have been compensated for by the ubiquitous consumption of vitamin C – mostly provided by sweet oranges.

More than sixty million tonnes of oranges are produced each year, two thirds of which are turned into juice, with Brazil, the United States, Mexico, India and China the world's top orange-producing nations. These good-looking sweet-tasting vitamin-enhancing fruits have therefore been transformed from once-exotic luxuries first cultivated in China into the modern world's premier healthy drink (see plate 12, top).

Banana

> **FAMILY: MUSACEAE**
> **SPECIES: MUSA ACUMINATA**
> **RANK: 98**

*The world's most popular tropical fruit which
now faces mass extinction*

BANANAS ARE SO popular to eat today because about 120
years ago they were identified by a number of American shipping
magnates as an ideal food for transportation from one part of the
world to another for the pursuit of profit. Their success depended
on farming the fruits cheaply enough and selling them to the mass
market in sufficient quantities – challenges that bananas have risen
to extremely well.

Until 1880 few people outside Asia and Africa ever ate bananas.
Initially cultivated in Indonesia, bananas had reached Africa by
the birth of Islam in *c.* AD 600. Muslim warriors then spread these
nutritious fruits to the Near East, Egypt and Spain as well as down
the coast of west Africa, where they were discovered by Portuguese
merchants in 1482. From here European explorers planted them on
the Canary Islands and took them to the Americas, seeding them on
the island of Hispaniola (now Haiti and the Dominican Republic)
in the early sixteenth century. Bananas were found to grow especially well in the tropical climate of countries such as Panama and
Mexico in Central America as well as throughout the islands of the
Caribbean.

Bananas belong to the tallest family of herbs in the world, sometimes reaching a height of ten metres. Their lack of wood means

they are completely frost intolerant, strictly limiting their range to tropical climes. The plants need between fourteen and twenty-three months of frost-free, sufficiently wet conditions to bear their fruit, which grows in bunches like long yellow up-pointing fingers. The name banana is itself derived from the Arabic word *banan*, which means 'finger'.

Despite their globalization, by the year 1900 bananas didn't seem set to change the world. Europeans and Americans were unsure what to make of the phallic-looking fruit, which no self-respecting lady could possibly peel and eat in public without first cutting it into pieces using a knife and fork. By the turn of the twentieth century they were still regarded by most people as rare and exotic. An article about bananas in *Scientific American* magazine in 1899 even included a list of instructions on how to eat them: '*The fruit is peeled by splitting the skin longitudinally and giving it a rotary motion with the hands . . .*'

Ten years later the fate of the banana had changed beyond all recognition. American shipping entrepreneurs pioneered a new business model by encouraging the growth of bananas on plantations in South and Central America so they could be transported in refrigerated steamships to ports such as New Orleans. From there they were packed on to ice-trains, 70 per cent of which went through the railway town of Fulton, Kentucky, where the main line divided into five tributaries, each of which headed out to different regions of the northern United States. The people of Fulton were kept busy servicing the steam trains and refreshing the wagons with ice. Keeping the fruits cool prevented them from ripening prematurely. So important was the banana trade to the people of Fulton that they marked each passing year with their annual banana festival – the first one staged in November 1963 – a mixed celebration of North, Central and South American culture.

Corporations such as the United Fruit Company, founded by Lorenzo Baker, a fishing captain from Boston, and the Standard

Fruit and Steamship Company, headquartered in New Orleans, Louisiana, found that bananas were perfect for marketing as an essential food. Of particular importance was the way that their ripening process could be postponed by refrigeration during transportation and then reactivated in heated rooms close to their point of sale. It was discovered that ethylene, a gas produced naturally by the fruit, could be used to accelerate the process, turning green bananas into a more visually attractive yellow.

American banana merchandizing companies successfully convinced everyday folk that life was incomplete without eating bananas. Everything about them was presented in ways that made bananas look and feel good: their thick skin was promoted as a natural protection against germs; the fruit was advocated as a source of vital vitamins; their convenience was emphasized in the form of a between-meals snack; and their soft texture made them ideal for mashing and feeding to small children. Best of all, bananas were 100 per cent safe to eat since domesticated varieties contained nothing as potentially hazardous as an easy-to-choke-on seed.

But man's dependence on a single mass-produced variety of seedless banana called Gros Michel (or Big Mike) very nearly ended capitalism's love affair with this golden-fingered fruit. A fungal wilt known as Panama disease (*Fusarium oxysporum*) first appeared in Asia in the early twentieth century. By the 1950s it had spread on the wind to the Americas, decimating the plantations that sustained the world's banana exports. Growers' businesses were wiped out (see also grape, page 309; potato, page 212; and apple, page 360). Even drowning crops in Bordeaux Mixture (see page 43) provided only limited protection.

By the 1960s Gros Michel bananas had become extinct. It was only thanks to the discovery of a new disease-resistant variety in Vietnam that the prospects for the genes of mass-produced bananas were revived. Cavendish bananas (*M. acuminata*) spread across the globe, and by the mid-1970s the banana business was back. Almost all bananas sold in shops and supermarkets throughout the world today are of the Cavendish variety. Connoisseurs say they taste less good than the once ubiquitous Gros Michel and traders agree that Cavendish bruise more easily in transport. But these are minor inconveniences, far outweighed by the enormous benefits of bananas with natural disease resistance.

But then about ten years ago, almost as predictably as the return of the Terminator in a Schwarzenegger movie, a new variant of the banana-decimating fungus began devouring its way through Cavendish crops in Asia. Experts predict its inevitable arrival in the banana-producing countries of Central and South America within twenty years. As yet there is no cure. A chronic lack of genetic diversity means that the chances of finding another disease-resistant substitute are remote. Genetic engineering, as with the apple disease fire-blight (see page 360), may be mankind's only escape route if it wishes to continue its love affair with the banana. Meanwhile, about seventy million tonnes of Cavendish bananas are consumed world-wide each year. For how long these cheap, affordable, plentiful fruits will retain their premier place rather depends on which side gains the upper hand in the ongoing tug-of-war between the natural and human worlds . . .

14

On Rivalry

How certain species have thrived with the rise of human civilizations, often despite attempts to keep them at bay.

Deer

> FAMILY: CERVIDAE
> SPECIES: CERVUS ELAPHUS
> RANK: 55

A fleet-footed, half-venerated, half-hunted herbivore that has succeeded through the constant pressure of natural selection

DEER ARE STEALTHY, herbivorous animals that currently boast healthy populations in every continent except uninhabitable Antarctica and Sub-Saharan Africa. The secrets of their present success were sown in the wild, millions of years ago.

Modern species of deer evolved from a common ancestor about twelve million years ago. The world then was full of wild predators including lions, lynxes, bears and wolves, just to name a few. The best survival strategy for these even-toed ungulates (ultimately related to camels, pigs, giraffes, sheep, goats and cattle) lay in speed and nimbleness. In such an environment the slowest to flee were those most likely to be caught by hungry predators, leaving the fastest, fittest individuals to live on and reproduce. With the passage of time, natural selection therefore saw to it that deer genes were honed to make this creature one of nature's most powerful long-distance runners, capable of speeds exceeding fifty miles per hour and of scaling heights greater than two metres.

The other trade secret belongs to the unique set of disposable antlers that are the hallmark of male European red deer (*Cervus elaphus*), usually called stags. A highly polygamous society underpinned by violent rutting contests ensures that only the strongest, fittest males earn the right to mate with a harem of up to twenty different females. This pattern of wild behaviour has also focused the deer's gene pool in favour of size and strength. Every year males grow a fresh set of antlers to display their current level of biological fitness. During the mating season males parade past each other comparing their body and antler sizes in a bid to determine which will have the right to mate. In the event they cannot decide, they lower their antlers and resort to a violent duel, a method that invariably sorts out which of them is fittest (see also sperm whale, page 173). Over thousands of generations such rituals have resulted in a super-race of strong swift creatures.

The stag's speed and strength have been enshrined in myths and legends that have come to characterize the relationship between these wild animals and humans over the last few thousand years. An association with the mystical god-like power of deer permeates many cultures. In the mythology of ancient Greece the stag was sacred to Artemis, goddess of the hunt, while Heracles was tasked with capturing the Ceryneian Hind, with its golden antlers and hooves of brass, in the third of his twelve epic labours. It was said this sacred creature could outrun an arrow.

The French nation owes a big debt to the magical powers of a female deer sent by God in answer to the prayers of Clovis I (AD 466–511), founder of the kingdom of the Franks, who was unable to find a way to lead his army across the river Vienne. A huge doe is said to have entered the river, showing Clovis and his men the best place to ford. As a result the Frankish army was able to defeat the Visigoths and their leader Alaric II in 507 at the battle of Vouillé, leading to Clovis's capture of south-western France and Toulouse.

By medieval times stags' antlers had become prized by the landed gentry as trophies, symbolic of man's power over nature. They still grace the halls, dining and billiard rooms in many of the stately homes and palaces of Europe. These were not just a Western fancy. Honda Tadakatsu (1548–1610), a samurai general, never once wounded in fifty-five battles, was famous for his lucky helmet, which was adorned with deer antlers. Honda's headgear possibly also reflected the traditional Far Eastern belief in the power of the stag's horns as an aphrodisiac. The soft velvet skin from a growing stag's horn, as well as the powder made from the crushed horns themselves, are still used in Chinese medicine today as a cure for impotence.

Musk deer (*Moschus moschiferus*), native to Tibet, China, Korea and Siberia, also have powerful sexual connotations. These deer have been hunted at least since the sixth century for the small gland positioned delicately between their genitals and tummy button. The world's most valuable perfumes are made from these glands, which are dried in the sun before being ground into powder. The resulting 'musk grain' is then dissolved in a tincture of alcohol to make perfume. By the beginning of the nineteenth century, musk grains from Tibetan deer were worth more than twice their weight in gold. Although synthetic substitutes are mostly used today, these deer are severely endangered because of over-hunting, despite a worldwide ban. A musk grain shipment of 700 kilos was seized in Japan in 1987, representing the harvesting of more than 100,000 deer.

Cousins of these deer, belonging to the family Cervidae, have no such problems. Not only have these creatures benefited from their natural endowments of speed and size, but their worldwide mythological associations have gripped the human imagination – to their considerable advantage. The sika deer (*Cervus nippon*) is a Far Eastern species closely related to the red deer, which has been transported all over the world to populate the ornamental deer parks of kings and aristocrats. Thanks to their unrivalled jumping skills, feral populations have since become widely established. Consequently this species is now common almost everywhere, from Britain to New Zealand and from Morocco to the United States.

Native populations of red deer, elk, caribou, moose and reindeer – all of which belong to the Cervidae – thrive today thanks to the demise of most of their wild predators at the hands of man. Wolves, jackals, bears and big cats – aggressive carnivores that present a threat to people – have either been eliminated, domesticated or sidelined so that wild well-adapted deer have few enemies left to check their growing numbers. Today's deer predators are mostly fast and metallic with dire consequences for all involved. More than 1.5 million car–deer collisions are reported in the United States each year, resulting in more than a billion dollars' worth of damage and the loss of 150 (human) lives. Meanwhile an estimated 13,000 moose have been killed in collisions with Norwegian trains since the year 2000. In other countries such as Argentina, deer are listed as a highly invasive species owing to their soaring populations and voracious appetite for young seedlings.

Deer have a long contradictory history with humans, with their status shifting variously from demi-god and ornamental marvel to prize game and vermin. Since neither life-form shows any natural limit in population size, the human–deer relationship is one potential flashpoint where the ongoing conflict between the evolutionary and ecological powers of man versus nature may find future expression.

Red Fox

> **FAMILY: CANIDAE**
> **SPECIES: VULPES VULPES**
> **RANK: 65**

A highly adaptable predator that continues to prosper despite man

WHAT'S THE BEST way to eat hedgehog? Most predators would think twice before assuming they were on to a tasty morsel when faced with a creature curled up into a tight defensive ball with razor-sharp spines pointing menacingly outwards. But not the fox, whose strategies for survival in the wild few other creatures on Earth can match. Calmly, it cocks its leg and squirts a jet of concentrated urine directly on to the hedgehog's back. As the animal recoils in shock its vulnerable underbelly is momentarily exposed, allowing the fox to go in for the kill.

Foxes are exceptional creatures. Over the last 2,500 years they have held their ground in the face of intense human encroachment by adapting, fleeing, double-crossing and manipulating their way to reproductive success. So great an impact have these creatures had on human culture that foxes have become worldwide icons in folklore, myths and legends from Europe to America, China, Korea and Japan.

Foxes are everywhere depicted as cunning. *Huli jing* are Chinese fox spirits that possess the bodies of beautiful young women in order to tempt and seduce married men. The Japanese fox spirit Kitsune, first recorded in AD 794, also assumes human forms. *Kitsunetsuki* refers literally to the state of being possessed by this fox spirit, which typically enters a young woman through her fingernails or breasts. Kumiho is a Korean fox spirit that transforms itself into the identical likeness of a bride at a wedding. No one can spot the change until the wedding night, when she finally takes off her clothes . . .

The fox's association with deception and guile also permeates Western culture, from Ben Jonson's *Volpone* ('The Fox'), the story of a man who tricks his friends and relatives into thinking he is dying so that they will bring him expensive gifts, to Joel Chandler Harris's *Tales of the Wily Br'er ('Brother') Fox*. Calling someone 'foxy' or a 'vixen' today is to suggest slyness or sexual mischief. This animal's distinctive reputation in human culture stems from the fact that foxes have, without deliberate encouragement from humans, prospered over the last 2,000 years while other similar species have either become dependent (domesticated dogs) or are in danger of extinction (from the European lynx to the grey wolf).

By far the most prolific fox species (there are twelve true fox species in all) is the red fox (*Vulpes vulpes*). These creatures have easily the largest worldwide distribution of any member of the dog family (Canidae) with an estimated range of seventy million square kilometres, nearly half the total landmass of the Earth. They thrive across all northern and eastern hemisphere habitats, with the exception of Greenland and Ireland. Populations in Europe stand at a healthy one million, with 250,000 in the UK, where they (nearly) all live in the wild.

The success of the red fox in the modern age is a triumph for natural selection. Without needing to be tamed as a pet or farmed as food for human consumption, this creature has been spectacularly effective at adapting to a world dominated by people without condescending to live by their rules. Perhaps such a spirit of defiance helps explains why, for nearly 1,000 years, humans have tried their level best to hunt foxes to oblivion using packs of specially bred well-trained domesticated foxhounds and servile horses (see page 246).

When William the Conqueror invaded Britain he brought with

him the European habit of hunting wild animals with dogs. By the sixteenth century foxes were the most widely hunted targets (rather than deer or wolves). Against all the odds (at least thirty hounds in a typical pack take up the chase) foxes have outwitted, out-manoeuvred and outrun their adversaries and survived in good numbers into the present day. Doubling back to confuse hounds, diving into water to obfuscate the trail of their scent and running over long distances at speeds of more than thirty miles per hour are characteristics of creatures determined not to lose their independence in the wild.

When humans began their ostentatious programme of building towns and cities on greenfield sites following World War I, foxes did not simply yield to the inevitable and flee into what was left of the woodland. They adapted by hiding in the day and feeding on the waste discarded by humans in urban areas at night. There are as many as 33,000 urban foxes in the UK (around 15 per cent of the overall population), which have migrated from their traditional woodland territories and now patrol leafy suburbs, raiding ponds for fish, bins for titbits and bird tables for seed. So flexible are the diets of these creatures that they can eat almost anything that lives, or has lived, from earthworms, which they feed to their pups, to snakes, rodents, rabbits, eggs, berries . . . and, of course, chickens.

Like cats and humans, foxes are popularly thought to enjoy killing for pleasure. Witness a fox raiding a chicken coop and no bird is spared, even though its killing spree usually results in far more dead meat than a single fox can possibly consume in one sitting. However, what humans may project on to the fox as a penchant for killing for fun is more likely a wild instinct that causes it to kill whatever meat is available in case of scarce future supplies. One by one it takes the chickens in its mouth, burying them in different locations (a technique called scatter-caching) to minimize the chances of another creature discovering the location of a single, large hoard. Chicken owners who discover their dead animals apparently discarded by the fox interrupt this process, hence the common perception that foxes, like people, kill for the thrill.

Perhaps foxes have been able to cope with adapting to a human-dominated world without compromising their wild ways because they are mostly solitary creatures. Each individual is used to fending for itself, unlike dogs or wolves, which hunt in packs for food. When myxomatosis struck the UK in 1953 (see page 347), decimating the rabbit population, foxes simply switched eating habits and hunted voles instead. In a fast-changing environment versatility, resource-fulness and independence of mind are characteristics favoured by natural selection, each generation becoming just a little more foxy than the last.

British fox-hunting only ever managed to slaughter up to 25 per cent of the wild fox population even when the sport was legal (it was outlawed in Scotland in 2002 and England and Wales in 2005). So successful has the fox been at maintaining its numbers – despite persecution – that the British gentry exported these creatures to the Americas and Australia to give them a decent adversary over-seas (see also rabbit, page 344). US Presidents George Washington and Thomas Jefferson both owned packs of foxhounds. Since then the American fox has been displaced by the European red fox. Meanwhile, owing to the lack of natural predators in Australia, special fox eradication programmes have recently been established to try to bring back thirty-one indigenous animal species from the brink of extinction.

There is a new predator though, which even foxes are unable to find a way of avoiding. Cars are thought to kill as many as 100,000 foxes on the roads of Britain each year, mostly at night. Natural selection has not yet taught foxes the Highway Code. Even though this new peril is not endangering the entire species, it is an unfore-seen consequence of modern motoring to control fox populations in ways that used to be accomplished by hunting with hounds. There are few better illustrations of the often cataclysmic collision between the systems of natural and artificial evolution than the sight of a bloody red fox splayed across a country road with its mechanical killer speeding off into the dark, oblivious of its deed.

Crow

> **FAMILY: CORVIDAE**
> **SPECIES: CORVUS BRACHYRHYNCHOS**
> **RANK: 64**

A global survivor that has recently taken up residence in towns and cities but which could inherit the Earth . . .

COULD OTHER ANIMALS one day become more intelligent than humans? The logic, according to the system of natural selection, isn't hard to grasp. Wild creatures best able to survive a world populated by seven billion humans will be those endowed with the highest levels of wit, brains and cunning that enable them to turn today's rapidly changing environments to their advantage. While it took humans about two million years to develop into intelligent beings with their own art, language and culture (see page 191), the challenges of living with humans today may mean other creatures develop similar ways of life a lot more quickly. Foxes represent such species on the land (see page 400) and crows lead the way among animals that fly.

Humans and crows evolved alongside each other in antiquity. Shamans, holy men from the hunter-gatherer world, painted images of crows in the caves where they contacted the spirit world. Some of these images, dating back 30,000 years, still remain in Lascaux, southern France. Ravens – larger, solitary members of the crow family – appear in legends contained in texts such as the *Epic of Gilgamesh* and the Jewish Torah, where their navigational wits were used by humans to divine for land. By the later Roman empire the association of crows with intelligence was well established. One of Greek writer Aesop's most famous fables tells how a crow found he was unable to reach the water in a pitcher. He solved the problem by dropping pebbles into the pitcher one by one until the water level rose sufficiently so he could quench his thirst. The moral, of course, is that intelligence, rather than brute force, is sometimes the surest means of survival.

Dramatic changes in the natural habitat since the end of the Roman empire haven't worried these creatures. On the contrary, like foxes, crows have adapted and prospered as a result of human population growth. Until medieval times their penchant for eating dead flesh was highly venerated in cultures throughout the world from Europe and America to the Far East and aboriginal Australia. Excarnation is the practice of leaving human remains in the wild to be eaten by birds and other scavengers. Crows and vultures, which picked at dead human flesh, were regarded by Neolithic pastoral people in Europe and the Near East as intermediaries with the spirit world, part of nature's endless cycle of birth, death and new life. Until the 1950s it was a tradition in Tibet to place the dissected remains of a loved one on a ceremonial altar for crows and ravens to carry off into the next life.

By medieval times, however, crows had become associated in the popular mind with disease, witchcraft and macabre death. Human bodies were frequently left to decay across the terrain of Europe and around the Mediterranean as Islamic and Christian civilizations fought for control of valuable overland trading routes, and crows gorged on these victims of warfare. Plague, the Black Death and famine afflicted Europe from the fourteenth to the nineteenth centuries and reinforced the association of these creatures with bad luck. Crows are an integral part of the opening scene involving witches depicted in William Shakespeare's great Scottish tragedy *Macbeth*.

About forty members of the crow family have been identified, ranging from the larger ravens to the smaller jackdaws and rooks.

On Rivalry

They are thought to have evolved from a common ancestor in Asia approximately five million years ago. As Charles Darwin deduced when he visited the Galapagos Islands, identical creatures when distributed over different, isolated habitats eventually develop small changes in anatomy and instinct that ultimately lead to the creation of related but genetically distinct species. When the Bering Strait was a swathe of coniferous forest approximately two million years ago, crows are thought to have migrated across to the Americas. The common crow, *Corvus brachyrhynchos*, has since become one of that continent's most successful avian species.

Survival for this native American crow has come from adapting its traditional meat-eating diet to the consumption of grain, in response to human farmers turning vast tracts of wild terrain into cultivated fields and pastures. Over the last 2,000 years crows, like foxes, have learned to eat almost anything. One recent survey found as many as 650 different types of food in the bellies of 2,118 crows. Their diet ranged from eggs, snakes and worms to noodles, chips and even fluorescent orange cheesy puffs.

As a result of urbanization in the late nineteenth and twentieth centuries, crows have adapted to living in every location from chimney stacks to the tops of advertising billboards. Walk down a typical city street today in most parts of the world and – apart from a few ornamental trees and perhaps a patch of grass – the only signs of nature will be crows. These are birds that have learned to live off mankind's most prolific by-product, urban refuse.

Reproductive success, flexible diet and adaptability of habitat all help a wild animal prosper in the modern world. But there is more to the crow's success. As naturalists spend time observing how crows live, the traditional perception of birds as dumb, instinctive creatures compared with mammals such as whales (see page 171), elephants (see page 175) or rats (see page 180) is undergoing significant revision. Bird brains, it is being discovered, may be wired very differently from those of mammals, but some species have at least an equal capacity for intelligence, communication and culture. The crow's brain relative to its body is among the largest in the animal kingdom, equivalent in size to that of a chimpanzee. They are even referred to by some experts today as flying monkeys.

Recently observed behaviour certainly supports that view. Crows, like humans and some primates, habitually manufacture tools to make feeding easier. American crows have been seen to craft splints out of wood to skewer insects lodged in crevices. Meanwhile, New Caledonian crows, challenged by problem-solving exercises in a research laboratory, were found to be capable of bending wires into hooks to retrieve a small bucket of food. On another occasion a raven in Maine, USA, untwisted ten separate wires, one by one, to open a squirrel-proof feeder.

Some ornithologists believe crows exhibit their own distinctive culture by acquiring and passing on knowledge, a skill normally credited only to humans. The recently observed antics of carrion crows living in Sendai, northern Japan, vividly illustrate this point. These crows wait for cars to stop at traffic lights, position nuts in front of the car wheels, and return to eat the kernels once the cars have moved off and crushed the shells. A similar pattern of behaviour has been found to have spread over a range of several miles during the last few years, demonstrating that these crows can pass on acquired knowledge through cultural exchange.

Are crows that treat cars as nut-crackers an example of creatures developing superior intelligence as a result of living in a modern world dominated by humans? Perhaps natural selection is now acting on crows, foxes and other non-human species to sharpen their intelligence, honing their survival skills in ways that may one day rival the ingenuity of mankind. A speculative zoological study on the possible course of evolution in a world after man lists crows as evidence that: '*intelligence as high as man's may evolve once more . . . After all, mammals were scurrying about the feet of dinosaurs for a good hundred million years before they came to anything.*'

On Rivalry

Birds, like mammals, survived the catastrophe that wiped out the dinosaurs (see page 126). Crows, once venerated by humans but now mostly regarded as scavenging menaces, could be man's intellectual successors in the event of another extraterrestrial strike.

Fruit Fly

ORDER: DIPTERA
SPECIES: DROSOPHILA MELANOGASTER
RANK: 15

*An ancient way of life that has prospered with
the rise of people and farming*

IF A NOBEL PRIZE for medicine could be awarded to a non-human species then a serious contender would have to be the common fruit fly. Thomas Morgan, an American Nobel Prize winner, was able to verify nineteenth-century monk Gregor Mendel's theory about genetics and inheritance only thanks to staring into the whites of a mutant fruit fly's eyes in 1910 (see plate 13, top).

In the year 1900, following the rediscovery of Mendel's pioneering work on pea plants, a race began between European and American scientists to discover the precise mechanism by which physical and behavioural features are passed on from one generation of living things to the next. The mechanics of inheritance had fascinated biologists even before Charles Darwin proposed his theory of evolution by natural selection in 1859. Morgan, professor of experimental zoology at Columbia University in Ohio, chose the fruit fly (*Drosophila melanogaster*) as a model species on which to base his research. This fly had a host of features that made it ideal for studying genetics. It was fast to breed, quick to incubate, easy to observe, a good size to handle and, best of all, because females are significantly larger than males, it was easy to distinguish one gender from the other – an important factor when embarking on a programme of selective breeding.

410

On Rivalry

Milk bottles were additional aids in Morgan's experiments, carried out inside his top-floor 'Fly Room', beginning in 1908. Working with his team of students, Morgan attempted to identify individual flies with some form of observable mutation. By breeding a mutant fly with an apparently normal partner he hoped to see how the mutation manifested itself in future generations. In 1910 Morgan struck the geneticist's equivalent of gold – a mutant male fly with white eyes.

Into the bottle went this precious creature, accompanied by a single female with standard-issue red eyes. Once nature had taken its course, Morgan could trace the pattern of the white eyes in the mutant's family tree as the gene cascaded down several generations. Before long Morgan and his students were tracing all kinds of different mutations – genes that gave flies pale yellow bodies, wrinkled hairs or short wings, each time confirming Mendel's theories concerning the pattern of genetic inheritance. Morgan's book *The Mechanism of Mendelian Heredity*, published in 1915, explained how genes responsible for different physiological traits were placed in fixed locations on sexually transferred chromosomes. At last biologists had a concrete system of inheritance to underpin Darwin's theory. What's more, Morgan's flies earned him the Nobel Prize for Medicine in 1933 for identifying chromosomes as the vector for the inheritance of genes.

So what? What's the big deal about scientists being able to spot genes in fruit flies responsible for white eyes? The true significance was revealed when it was discovered that humans and flies share many of the same genes. These include the crucial genes responsible for embryonic body design as well as many that cause hereditary disease. Up to 75 per cent of known human disease genes are similar or identical to those found in fruit flies, allowing scientists to explore potential treatments for human afflictions such as Alzheimer's, Huntington's and Parkinson's simply by experimenting on flies. Such research has led to yet more scientists winning Nobel Prizes, thanks to their work on easy-to-propagate fruit flies.

If fruit flies sound like an unlikely source of clues leading to cures for disease, then maggots (fly larvae) as a way of combating antibiotic-resistant bacteria such as MRSA will probably sound even more far-fetched. However, before the development of antibiotics in the mid-1940s (see page 328) one of the most successful treatments for infected open wounds was the deliberate placing of maggots on diseased flesh. Maggot biotherapy was originally deployed by the Aborigines in Australia and the Mayans in Mexico, yet the practice was rediscovered in the West only during Napoleonic times and used later during the American Civil War (1861–5). Confederate medic Dr J. F. Zacharias wrote about the miraculous healing powers of maggots, which '*in a single day would clean a wound much better than the agents we had at our command*'. During World War I it was found that if the open wounds of soldiers left on the battlefield became infested with maggots they remained infection-free, typically healing within six weeks.

Live maggots kill bacteria by secreting their own antibiotic juice. They also feed solely on dead tissue, leaving healthy flesh intact. So effective are maggots at killing bacteria that larvae from the greenbottle fly species *Phaenicia sericata* are now being used in place of penicillin and other antibiotics to fight modern resistant bacterial strains such as MRSA. Maggot medicine, now in use in as many as twenty countries, received approval from US and UK medical authorities as recently as 2004 as an alternative to antibiotics. Maggot-laced dressings, despite being a little tickly for the patient, have one other major advantage over antibiotic drugs: they do not encourage the selection of mutant antibiotic-resistant bacterial strains.

Positive feelings about the impact of everyday flies, past and present, go against normal human sensibilities. Since biblical times various cultures have regarded flies as evasive insects that congregate only around excrement and rotting flesh. Often they have been depicted as incarnations of death and the devil. Beelzebub, Lord of the Flies, took his place alongside Lucifer and Leviathan next to Satan in a seventeenth-century French Catholic witch-hunter's depiction of Hell.

Mythological omens in the minds of many turned into grim reality shortly before the turn of the twentieth century. During the Spanish–Mexican War (1898) it was confirmed by a US government report that flies were the most likely transmission route for the typhoid bacterium, a major cause of death of American soldiers in the four-month-long conflict. Today more than one hundred diseases are known to be transmitted by flies to humans. Their maggots hatch from eggs laid in animal dung, itself infested with pathogenic bacteria, viruses, protozoa and other disease-causing agents. The maggots pupate and metamorphose in adults, which transfer these disease organisms to humans when they walk on their food, or regurgitate digestive saliva on it. One type of fly (*Muscina stabulans*) uses its antennae to detect buried bodies (human or animal) on top of which it lays its eggs. Once hatched, its larvae instinctively burrow down and invade the corpse, feeding on the rotting flesh. Forensic scientists sometimes use the presence (or absence) of maggots like these to determine the time and location of death.

Human population booms and the conversion of land into pasture for dung-dolloping farm animals have ensured that everyday flies, of which there are thought to be about 240,000 species, continue to thrive all over the world. Their evolution since the age of dinosaurs – amber fossils containing trapped flies have been found dating back to the mid-Triassic Period – continues unabated today. While bloodsuckers such as mosquitoes (see page 130) and the tsetse fly (see page 138) have had their own specific impact, the Nobel-Prize-winning fruit fly *Drosophila*, adopted by researchers in their dedicated fly rooms, earns its rightful place alongside those species that have most changed the world. It qualifies not just by virtue of the genetic insights it has provided, but also for its duplicitous role as both transmitter of and cure for diseases in humans.

Tick

FAMILY: IXODIDAE
SPECIES: IXODES SCAPULARIS
RANK: 45

*A miniature arthropod that used to feed on dinosaur
blood and is thriving today owing to climate change*

IT IS TEMPTING to think that the most severe effects of global warming are still some way off – at least on a human timescale. Perhaps it is inevitable that at some point there will be no polar ice caps and sea levels will have risen by between five and twenty-five metres. But most scientists predict that this scenario cannot realistically happen until towards the end of the twenty-first century at the earliest. And an extreme event such as the imminent melting of the Wilkins Ice Shelf, an ice sheet the size of Jamaica which is currently breaking off Antarctica, although it sounds dramatic, won't really affect sea levels since most of this ice is already floating in water.

But there is another urgent reason why global warming matters today. Small bloodsucking arthropods called ticks are highly sensitive to changes in atmospheric temperature. Because of global warming and the fact that winters all over the world are getting milder, tick numbers are currently booming and their range is extending northwards. These insects harbour a number of dangerous diseases, which are now sweeping through the human populations of North America, Europe and Asia on an epidemic scale.

Many people are ignorant of the hazards of tick bites. It was only relatively recently that Swiss-born American scientist Willy Burgdorfer first identified the type of bacterium that causes Lyme disease, an infection transmitted by ticks to humans. This pathogen, discovered in 1982 and named *Borrelia burgdorferi* in honour of its discoverer, is a missile-shaped spirochaete that swims around in the bloodstream propelled by a coil of twisting tail-like flagella. These Prokaryotic organisms, related to the very earliest living cells, hide away from the oxygenated atmosphere by seeking refuge in the bodies of multicellular creatures be they reptiles, birds, mammals, humans – or ticks.

Since Burgdorfer's discovery, a host of frequently recurring symptoms such as fever, headache, fatigue, depression and arthritis have been diagnosed as Lyme disease. If left untreated the spirochaetes reproduce in the soft tissues, sometimes causing irreversible damage to eyes, heart and brain. Although few individuals are known to have died from this disease, the quality of life in many sufferers has been severely compromised. Thousands of new cases of Lyme disease are being diagnosed each year in North America and Europe, making this infection one of the fastest-emerging illnesses in the world.

Ticks are arachnids (the class that includes spiders), arthropods that evolved more than a hundred million years ago, initially feeding opportunistically on the blood of dinosaurs. They have thrived in the increasingly grassy world that has emerged in the last fifty million years with its vast populations of grazing mammals. Different tick species have different animal hosts, the most popular today being deer (see page 396) and sheep (see page 224). After several days gorging on the blood of an animal, the tick loosens its grip and falls into the long grass, where the female lays up to 10,000 eggs. These hatch in the springtime, waiting for an unsuspecting animal to brush past, allowing the baby tick to latch on and gorge for itself. Disease-causing pathogens pass from the tick's saliva into the bloodstream of its host.

Lyme disease is not the only potential check on human well-being transported by hard-bodied arthropods such as *Ixodes scapularis*. Other types of bacteria, viruses and protozoa are similarly well served by these tiny bloodsucking vectors. Rocky Mountain spotted fever is an American disease spread by ticks that range from Peru, throughout the United States and into Canada. The disease is notoriously hard to diagnose and if left untreated can be fatal in up to 5 per cent of cases. *Rickettsia rickettsii*, the offending bacterium, lives in the male tick's semen and is passed on to the female during copulation. The bacterium then infects her eggs. Once hatched, whatever creature the baby ticks bite also becomes infected by the disease, making the tick population a persistent bacterial reservoir.

Another disease, tick-born encephalitis (TBE), is a viral pathogen that attacks the central nervous system. Although preventable by vaccination, TBE can cause crippling neurological damage in unprotected people. Up to 20 per cent of those infected are left with lifelong complications, sometimes losing their ability to walk. As many as 12,000 cases are reported in Europe and Russia each year.

The impact of these hard-to-diagnose tricky-to-treat diseases is only just beginning to be felt. Ticks die off in harsh winter climates, but owing to a succession of mild winters their range is increasing, especially in highly populated areas of the northern hemisphere. People living or holidaying on the 220,000 islands off the east coast of Sweden have to endure one of the most tick-infested environments in the world. Since records of viral tick-borne infections began in 2004, cases have increased from 224 to 584 in just four years, despite the increasing availability of vaccines. Recent research suggests these infections are now spreading to western parts of the country.

'Boat clinics' visit the islands offering free vaccinations against TBE, and holidaymakers are told to be far more concerned about the hazards of tick bites than of sunburn. Meanwhile, a third of Swedish ticks also carry Lyme disease (which, though treatable, has no vaccine) and cause infections in as many as 10,000 people a year – a big number in a country with a population of just nine million.

My eldest daughter contracted Lyme disease when she was five years old. The tick must have brushed on to her from the long grass at the end of our garden. We knew nothing about the hazards of tick bites at the time, but when the telltale bullseye rash appeared it was clear that spirochaete infection was a possibility. Several blood tests later and the diagnosis was confirmed. Fortunately catching it early on meant an eight-week course of antibiotics stopped these spiralling bacteria in their tracks.

Local doctors were dumbfounded. Incidents of Lyme disease in south-east England were almost unheard of. Not any more. As the viable habitat for tiny bloodsucking creatures creeps ever northwards, people who live in parts of the world unused to diseases traditionally called 'tropical' will have to reassess the risks of walking through innocent-looking countryside or woodland. The arrival of disease-infected ticks is among the early-warning signs of a very different world, soon to come.

Lactobacillus

> **CLASS: BACILLI**
> **GENUS: LACTOBACILLUS**
> **RANK: 5**

Gutsy bacteria that have, in their own small way,
profoundly boosted the fortunes of man

ASTRONOMERS LIKE TO think BIG. While the number of galaxies in the observable universe is thought to be in the region of about a hundred billion, the number of stars (like our Sun) in our galaxy, the Milky Way, is at least twice that number. Actually, those who enjoy *really big* numbers are better served by looking inwards rather than outwards. Take the number of living cells in an average human body – approximately ten trillion. Even more spectacular is the number of bacteria that live inside the human gut – *ten times as many as there are cells in the body* . . . In terms of discrete living components, you and I are ten times more bacterium than human. It is only reasonable that the historic contributions of such organisms should be well understood.

Perhaps it is because of Louis Pasteur's revelation of the role of bacteria in disease or Alexander Fleming's discovery of antibiotics (see page 326) that most people today associate bacteria with a negative impact on humanity. That is unfair – witness the contribution of cyanobacteria (see page 20), which gave birth to atmospheric oxygen, or *Rhizobia*, which fix nitrogen in the air to nourish plants (see page 30). Indeed the evolution of all higher forms of life is essentially a by-product of bacterial processes. The same as goes on inside our guts night and day.

The huge growth in the human population over the last 2,000 years – along with its communities of billions of sheep (see page 224), cattle (see page 228), pigs (see page 221) and other farm animals – has provided a huge boost for bacteria that survive only in oxygen-free environments. People (and animals) are excellent hosts for such bacteria. Not only do our intestines provide a safe oxygen-free habitat for them but they also get very well fed. It should come as little surprise then that many bacteria that have made people and animals their home also seek to improve their adopted host's health.

Such were the thoughts of Ilya Ilyich Mechnikov (1845–1916), a Russian microbiologist best remembered for discovering the process of phagocytosis in the human immune system (see page 37), for which he won a Nobel Prize in 1908. During his research he began to suspect that certain bacteria that live inside the intestines are responsible for human fitness and longevity. He was especially intrigued by a genus of bacterium called *Lactobacillus*, which converts lactose (a sugar found in dairy products) into lactic acid. Until the end of his life Mechnikov was convinced these micro-organisms held the key to healthy living, so he wrote a book about it called *The Prolongation of Life: Optimistic Studies*.

Minoru Shirota (1899–1982), a Japanese scientist, picked up on Mechnikov's ideas a few years after the Russian's death. During the 1930s he experimented by mixing a new strain of *Lactobacillus* with skimmed milk to make a probiotic drink that he called Yakult. His aim was to stimulate the cultivation of healthy bacteria inside the gut to prevent infection and boost the immune system – vital if Japanese children of the time were to avoid malnutrition and disease. When Yakult came on to the market in 1935, the drink was an instant hit, especially in Japan and Taiwan. It is still sold in more than thirty countries today.

Strains of *Lactobacillus* that live inside the colon perform at least two functions that have been of great assistance to humans through-out history. First they help compensate for lactose intolerance, a genetic deficiency in many people. *Lactobacillus* is responsible for the fermentation process that turns milk into cheese, breaking down the sugars in milk (lactase) into digestible lactic acid. This process continues inside the human gut, helping humans tolerate a milk-based diet. Over time, such symbiotic assistance has become incorporated into the genomes of many people in milk-drinking cultures. A second benefit of *Lactobacillus* is that the process of creating lactic acid protects against less beneficial bacteria and other microbial pathogens (viruses, protozoa or fungi) that cannot thrive in acidic environments (see plate 16, bottom). The presence of lactic-acid-producing *Lactobacillus* is therefore one of the body's defences against infections spreading via the gut. It is also what stops fermented cheese from going off quickly. Indeed, were it not for the natural preservative effects of a strain of bacterium called *Lactobacillus plantarum*, which occurs naturally in plants and fruits, merchants in ancient Greece would never have been able to trade their precious crops of olives around the Mediterranean without them turning sour in the sun (see page 215).

The impact of bacteria on human history is woefully underwritten, healthy bacteria particularly. Pathogenic diseases caused by other types of bacterium (such as typhus) have left far more tangible evidence in their wake than the therapeutic benefits of probiotic good health. Yet boosts to immunity, higher acidity levels, lactose tolerance and food preservation make an impressive list of benefits. It is ironic to think that these bacteria may have been just as instrumental in supporting the massive rise in the human population as Alexander Fleming's bacteria-snuffing penicillin.

Herpes

> FAMILY: HERPESVIRIDAE
> SPECIES: HERPES SIMPLEXVIRUS
> RANK: 47

A highly infectious genetic agent that has colonized
most of the human population

VIRUSES HAVE HAD a fundamental yet little-understood impact on the history of life on Earth and on humanity itself, as the decoding of the human genome has shown. Despite the eradication of smallpox (see page 13) or the present containment of avian influenza (see page 5), viral diseases are continually reshaping the lives of countless individuals, nations and cultures. It is therefore appropriate to end our survey of one hundred species that have changed the world with a virus family that is ubiquitous in all humanity – past and present. The name 'herpes' comes from the Greek *erpis*, meaning 'to crawl', so-called by the father of European medicine Hippocrates (460–370 BC). It was a good name for a disease in which labial blisters and cold sores creep across the surface of infected human flesh.

Eight types of herpes virus infect humans, ranging from chicken-pox and shingles to glandular fever (Epstein-Barr virus). The two most significant are the closely related *Herpes simplex* virus (HSV) 1 and 2. The first lives inside the mouth, causing cold sores around the lips, while the second causes similar blistering to the genitalia. These two viruses are not limited to any particular region, race or gender. HSV 1 – the oral form – is estimated to be carried by a staggering 70 per cent of the human population – that's more than *four billion* individuals. Although not all infected people develop symptoms, each one is a carrier. As with HIV/AIDS (see page 7), humans infected with herpes carry the virus for life. Normally, symptoms recede after several weeks, and the virus retreats into the sensory ganglia in tissues around the ear, where it lies in a latent state, unable to be detected by the human immune system. Occasionally (for reasons not yet fully understood) it rears back into life, hijacking cells and reproducing its own viral matter to be spread through saliva, mucus or pus.

While most humans harbour herpes in a latent form, the infection does not generally appear to be fatal (except occasionally in some individuals, where it can attack the central nervous system causing encephalitis). However, if the human herpes virus happens to skip across the species barrier to infect, say, a gibbon, it is almost always lethal. Similarly, when the Simian Herpes B virus, common in macaque monkeys, infects a human there is a 70 per cent chance of death.

Herpes is a typical example of how viruses can establish a kind of evolutionary equilibrium with their host species. But if the host changes, the consequences for the evolution of other species can be catastrophic. It is even thought by some experts that a cause of the demise of the Neanderthal species of humans, the last to survive apart from *Homo sapiens*, was a leap of the herpes virus across the species barrier from *Homo sapiens* to *Homo neanderthalensis*. Such an interpretation would explain the selective demise of that single species while leaving *H. sapiens* unharmed. This theory is supported by the fact that there is so little evidence of interbreeding between the *H. sapiens* and *H. neanderthalensis* genomes.

While the spread of oral or genital herpes throughout the human population is not presently regarded as a major health threat, there are ominous signs that this irremovable incurable virus may carry a nasty sting in its tail. Recent research suggests that the exponential rise in Alzheimer's disease (or senile dementia) may be due to HSV

1 and 2 attacking people who have reduced immunity in old age. When the virus wakes up, as it often does, it is able to penetrate the central nervous system and enter the brain. Here it wreaks havoc, damaging brain cells (see plate 16, top), which degenerate, leading to mood swings, memory loss, a breakdown of communication skills, disorientation and eventually death. Alzheimer's disease (first identified in 1906 by German psychiatrist Alois Alzheimer) is currently the most expensive disease to treat in the developed world. The annual costs of care are estimated to run to $100 billion in the United States alone. Approximately twenty-six million people suffer dementia worldwide. Owing to the world's ageing population the number is expected to rise dramatically by 2050.

There are no known strategies for vaccinating against persistent viruses such as herpes and HIV. Such viruses represent the front line in humanity's struggle to secure control of the evolutionary process for its own ends – whether for food production, material wealth, longevity and recreation, companionship or in the pursuit of beauty. Another way of picturing this contest is to travel full circle, back to the beginning of life itself. All life, ever since its original incarnation roughly four billion years ago, can be viewed as an ongoing struggle for evolutionary supremacy between loose strands of genetic code (viruses) and cellular forms of life (bacteria). Both have continued their quest for immortality through higher forms of life from the leafy tissues of plants to the bloody bodies of creatures. As part of this conflict, elaborate survival strategies have evolved, such as the biodiversity of sexual reproduction (see page 41) or the gene-splicing systems mastered by genomic retroviruses.

Humanity's attempt over the last 12,000 years to craft nature to suit its purposes is superimposed uneasily on this age-old struggle. Exposure to viruses may therefore present the greatest threat to mankind and its precariously balanced artificial construct. As humans demonstrated so effectively with the release of myxomatosis in 1950, a virus can systematically all but wipe out a single species, while leaving an almost identical genetic line unscathed (see page 347).

What if such an event were to strike humanity? Would a world without people return to the state of nature as it was four million years ago before the first *Australopithecus* pivoted up on to two feet? Or would some other form of intelligent life eventually evolve and similarly try to superimpose its own rational way of thinking on to the self-organizing systems of nature? Who knows what on Earth would evolve next?

The Ladder of Life

*The top 100 species ranked
in order of overall impact*

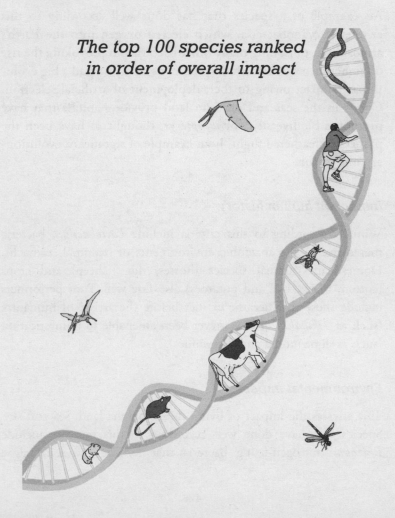

WHICH LIVING THINGS within our survey of '100 species that changed the world' have had most overall impact? The table on pages 436–9 presents one interpretation. Five simple (but entirely arbitrary) criteria, worth 100 points each, have been devised to assess the impact of each selected species.

Evolutionary impact

An example of a species that has done well according to this criteria is cyanobacteria, which ejected oxygen into the Earth's atmosphere, poisoning many other bacteria but provoking the rise of higher forms of life. Modern humans have also had a big evolutionary impact owing to their deployment of artificial selection. Corals in the seas and oaks on land provide habitats that have promoted biodiversity. *Archaeopteryx*, thought to have been the pioneer of feathered flight, is an example of significant evolutionary innovation.

Impact on human history

Winners according to this criteria include *Lactobacillus*, bacteria that thrive in the anaerobic environments of mammal stomachs. Domesticated animal species (horses, dogs, sheep) and crops (cotton, maize, rice and potatoes) also fare well. Poor performers include those that became extinct before the arrival of humanity (such as *Tiktaalik*) or have never been amenable to domestication (such as slime mould and dragonflies).

Environmental impact

This assesses the impact of living things on the land, sea and sky. Species that have done well according to this measure include *Rhizobia* (nitrogen-fixing bacteria that fertilize the soil), algae (absorbs CO_2 and emits O_2), cows (which emit methane) and cotton (responsible for drying out the Aral Sea). Species with little discernible contribution include bats, olives, Rhyniophytes, HIV/AIDS and potyvirus.

Global reach

This score is determined from the extent of a species' range when at its height. Those that score well have dominated the land (such as *Lystrosaurus*, smallpox, earthworm), the air (*Quetzalcoatlus*) or the seas (sperm whale). Those that fare less well are local niche growers such as durian and vanilla.

Longevity

Even extinct species such as *Lepidodendron* can score highly (150 million years compares very well against the 160,000 years of *Homo sapiens*), whereas more recent-to-evolve species – especially those domesticated by and dependent on humans for their survival – fare worse.

The top spot ...

The humble earthworm takes pride of place in the top 100 with 431/500 points, even though it doesn't come top in any single category. It is a suitable winner, being described by Charles Darwin in memorably glowing terms: '*It may be doubted whether there are many other animals which have played so important a part in the history of the world, as have these lowly organized creatures.*' (See page 93.) Agriculturally based human civilizations could never have emerged without the constant recycling of these subterranean wrigglers, which also rank high up on the list of nature's most robust, enduring and environmentally beneficial creatures.

What about humans?

Our species, *Homo sapiens*, comes sixth, and is the only mammal featured in the top 10. We lose out chiefly as a result of our recent evolutionary emergence. Human ancestors also fare well because of crucial evolutionary innovations such as increasing brain size (*Homo erectus*) and learning how to stand up and walk on two feet (*Australopithecus*). Without these the platform for artificial selection, which has propelled our species *Homo sapiens* so high up the table, might never have become established.

The top 10

Others in the top 10 divide into two groups. The first are those that emerged early on in the evolution of life on Earth and have contributed dramatically to the environment, such as cyanobacteria, algae, *Rhizobia* and stony corals. The second are those that have made a profound difference to human history, such as *Penicillium*, yeast, *Lactobacillus* and influenza.

Other high achievers

Thanks to the variety of the five criteria, extinction is no barrier to success; witness the *Lepidodendron*, that pioneering tree of the Carboniferous Period that left so much fine-quality coal for human use. Meanwhile, the survival skills of eucalyptus, with its recent global spread, utility to man and devastation of other species, are reflected in its elevated position, as are the environmental changes wrought by cows, predominantly in terms of bolstering human population levels and as a premier source of greenhouse gas emissions.

The Ladder of Life

Group selection

Each species in the table has been categorized into one of twelve groups: virus, bacterium, fungus, protoctist, invertebrate, fish, non-flowering plant, flowering plant, amphibian, reptile, bird and mammal. By far the most dominant and influential evolutionary force on Earth, according to this analysis, are the flowering plants (with 32 entries), followed by mammals (17), invertebrates (16), fish (6), viruses (5), non-flowering plants (5), bacteria (5), fungi (4), reptiles (4), birds (3), protoctists (2) and amphibians (1). Despite their predominance in the top 100, there are no flowering plants at all in the top 10 species, suggesting that flowering plants, although the most dominant life-force, are more the foot soldiers of evolutionary change than its pioneers.

Domestic success

The chances of making it into the top 50 are substantially greater for species whose genes have responded vigorously to the demands of artificial selection. Sheep, rice, sugarcane, maize, cotton, dogs, potatoes, cats and pigs are, as a result, global players in the few super-species that have come to dominate life on Earth.

Extinction still no barrier

Despite their demise approximately 250 million years ago trilobites (ranked 31, see also page 58) were once so dominant that they are among the most common fossils found in the world today. Their pioneering system for predation (underwater vision) was what probably triggered such extraordinary success. The genes of *Tyrannosaurus*, just three places behind at number 34, created the ultimate theropod, a lineage that lasted from the first appearance of dinosaurs 230 million years ago through to their demise approximately 170 million years later. Birds, such as *Archaeopteryx* (ranked 41) are theropod offshoots, so the *Tyrannosaurus*'s design lives on in the world's favourite food, chicken (see page 232), albeit in a much reduced and modified form. *Quetzalcoatlus* just makes it into the top 50, representing the first flying vertebrates, the pterodactyls – creatures able to circumnavigate the world.

Top trumps

How come elephants are so high up the table? After all, these are creatures limited in range (Africa and Asia) and with markedly less impact on human history compared to, say, other transporters such as the horse (ranked 58) or camel (ranked 87). On reflection, however, it is their ecological contribution that propels these majestic creatures up the ladder. Before the arrival of Neolithic farming approximately 12,000 years ago, these were nature's premier lumberjacks, along with fire, responsible for clearing forests in ways that allowed vital new growth. At the other end of the scale their dung provided essential habitats for seeding the nests of termites, creatures essential for the process of decomposing wood. Horses and camels, both beneficiaries of human domestication, have far more limited environmental credentials.

The biggest loser?

It must be the dragonfly, in at number 50. If this table reflected life on Earth 300 million years ago, there is little doubt that these aggressive carnivorous creatures would have made it into the top 10. Their declining relevance is reflected in their smaller size and the ecological niches in which they live. However, they are still successful, unlike many creatures of that era, such as the pioneering amphibian *Eryops*, which became extinct aeons ago and therefore doesn't make it into today's top 100.

The Ladder of Life

Just slipping out of the top 50

The lobe-finned fish would have featured more highly in the table had it been drawn up in the pre-human era, but today only a few lobe-fins survive (for example, the beautiful blue coelacanth, which was 'rediscovered' in a South African fishing catch in 1933). Lobe-fins cling on at the margins of the oceanic ecosystem, hardly reflective of their momentous evolutionary importance as the lineage that developed much of the apparatus required for fish to learn to live on land (see page 106).

Why aren't horses higher?

Surely horses could do better than a mere 58? After all, they have dominated the history of humanity for the last 4,000 years from being mankind's pre-eminent weapons system to its most ubiquitous form of transportation. Yet, despite their impact on human history (where they rank sixth) domestic horses have had little if any environmental impact. Being a domesticated species (wild horses became extinct several hundred years ago) their longevity score inevitably suffers. Horses are not generally farmed for consumption, putting the populations of cows (which provide meat and transport) and sheep (a source of wool and meat) far higher in the human-centric world. Finally, the age of the horse as premier battle system came to an end with the rise of the petrol engine, putting these creatures into a group of those falling into steady decline.

A tale of two opposites

You couldn't get two species much more different than *Lystrosaurus* (an extinct early mammal) and tobacco (a flowering plant), yet they cosy up alongside each other at numbers 66 and 67. The former was a lifeline for the lineage of vertebrate land-life that evolved to become mammals by surviving the perilous Permian Extinction 252 million years ago. The latter is a plant that has been responsible for the death of hundreds of millions, if not billions, of humans since it was globalized by European explorers who 'discovered' the Americas in the late fifteenth century. Both have had dramatic impacts on life as it is today, but for completely different reasons.

Sounds like sense

Bats, the world's only flying mammals, squeaked into the top 75 thanks to their extraordinary capability to see in the dark using sound waves. This survival edge is matched by their unrivalled diversity (there are more than 1,000 species, representing 20 per cent of all mammal species), meaning they are well equipped for future turmoil if some virus or environmental catastrophe unfolds. My bet is that in some future table bats might rise even higher.

Flower power

Dominating the lower ranks of the top 100 species are the flowering plants that have transformed human history, and with it the planet and other life. Material wealth comes from rubber trees; health from cinchona; mind numbing from poppies and coca; gastronomic pleasure from olives, cacao, oranges and bananas; while the lotus and lavender represent symmetrical beauty and perfumed fragrance. Evolutionary ingenuity and genetic pliability combine at their absolute best, it seems, in species rooted to the ground – perhaps because they rely most on other species for services such as pollination and spreading seeds while having evolved an awesome array of chemical defences against being eaten.

The Ladder of Life

Fastest riser

Conduct this survey just thirty years ago and HIV/AIDS would have been nowhere to be found because, at least in terms of human awareness, it didn't even exist. Yet here it is, powering into seventy-seventh place. There is still no cure for this retrovirus (see page 7), although a delay in the deadly effects of HIV has been accomplished using antiviral drugs. Worse is that the rate and extent of human infection is probably dramatically underestimated worldwide, and the long-term effects of what this potentially species-changing disease may do to human populations have yet to play out.

Future prospects

Of all the species in the top 100 there is just one that earns a place for its *potential* impact. Pseudomonas (see page 26) makes it only because of its ability to create biodegradable plastic out of artificial pollution. How quickly such alchemy could contribute towards reducing the enormous problem of non-biodegradable plastic waste that pollutes the oceans is as much dependent on humans as on bacteria. But the potential for organic solutions to the material needs of humans is so enormous that this species scores highly in terms of environmental impact just by being possible.

The bottom spot

Being among the top 100 is what really counts, but as in any league table something has to take bottom place. The potyvirus squeezes in after provoking the first financial crash in modern history and in the process propelling its host species, the tulip, to pre-eminence in the human-dominated world. Without other notable accomplishments, this virus languishes down at number 100, and I would not be surprised if, in the event of future (or rival) surveys, it fell out altogether.

What on Earth Evolved? . . . in Brief

OVERALL RANK	SPECIES	EVOLUTIONARY IMPACT		IMPACT ON HUMAN HISTORY		ENVIRONMENTAL IMPACT		GLOBAL REACH		LONGEVITY		TOTAL SCORE	GROUP	PAGE NUMBER
		SCORE	RANK	SCORE	RANK	SCORE	RANK	SCORE	RANK	SCORE	RANK			
1	EARTHWORM	87	17	93	8	95	4	82	11	74	16	431	Invertebrate	93
2	ALGAE	95	3	41	68	89	7	100	2	95	2	420	Non-flowering plant	44
3	CYANOBACTERIA	100	1	13	83	100	1	90	6	100	1	403	Bacterium	20
4	RHIZOBIA	65	28	85	25	87	8	74	36	85	5	396	Bacterium	30
5	LACTOBACILLUS	74	24	89	17	56	22	100	1	76	13	395	Bacterium	419
6	HOMO SAPIENS	98	2	100	1	98	2	83	10	14	78	393	Mammal	191
7	STONY CORAL	91	13	74	40	85	10	63	56	78	11	391	Fish	52
8	YEAST	69	27	92	11	32	42	100	3	82	7	375	Fungus	302
9	INFLUENZA	42	55	90	16	34	35	100	4	85	6	351	Virus	4
10	PENICILLIUM	93	8	95	3	13	65	60	60	75	15	336	Fungus	326
11	LEPIDODENDRON	65	29	74	41	68	14	80	17	45	32	332	Non-flowering plant	82
12	SEA SQUIRT	91	14	11	100	76	12	72	37	80	8	330	Fish	68
13	HOMO ERECTUS	84	19	92	12	67	16	66	47	20	74	329	Mammal	187
14	MOSQUITO	85	18	82	29	23	54	72	38	66	22	328	Invertebrate	130
15	FRUIT FLY	62	34	89	18	23	55	82	12	68	20	324	Invertebrate	410
16	EUCALYPTUS	45	50	73	44	87	9	76	28	42	36	323	Flowering plant	271
17	COW	40	56	89	19	98	3	78	24	7	86	312	Mammal	228
18	FLEA	78	22	82	30	13	66	76	29	62	26	311	Invertebrate	134
19	WHEAT	60	36	92	13	59	21	70	41	24	67	305	Flowering plant	202
20	ROUNDWORM	80	21	13	84	35	34	95	5	80	9	303	Invertebrate	55
21	OAK	72	25	84	26	30	47	66	48	45	33	297	Flowering plant	142
22	NORWAY SPRUCE	28	70	69	49	64	20	68	43	68	21	297	Non-flowering plant	90
23	AUSTRALOPITHECUS	93	9	94	4	54	25	32	79	23	71	296	Mammal	184
24	GRASS	35	63	67	52	67	17	85	9	37	41	291	Flowering plant	381
25	ANT	55	40	43	65	45	28	82	13	66	23	291	Invertebrate	161

The Top 100 Species

OVERALL RANK	SPECIES	EVOLUTIONARY IMPACT		IMPACT ON HUMAN HISTORY		ENVIRONMENTAL IMPACT		GLOBAL REACH		LONGEVITY		TOTAL SCORE	GROUP	PAGE NUMBER
		SCORE	RANK	SCORE	RANK	SCORE	RANK	SCORE	RANK	SCORE	RANK			
26	SHEEP	40	57	94	5	68	15	78	25	8	84	288	Mammal	224
27	RICE	50	46	93	9	65	19	64	55	7	87	279	Flowering plant	236
28	RABBIT	35	64	64	54	77	11	68	44	33	53	277	Mammal	344
29	SUGARCANE	55	41	93	10	55	24	62	58	4	95	269	Flowering plant	206
30	WATER MOULD	25	71	87	23	44	31	71	40	40	38	267	Protoctist	40
31	TRILOBITE	94	5	13	85	13	67	78	26	69	19	267	Invertebrate	58
32	RAT	46	49	82	31	24	50	80	18	33	54	265	Mammal	180
33	HONEY BEE	60	37	91	14	25	49	42	71	44	34	262	Invertebrate	157
34	TYRANNOSAURUS	70	26	13	86	33	40	82	14	61	29	259	Reptile	125
35	MAIZE	50	47	88	21	48	26	66	49	6	91	258	Flowering plant	241
36	ACACIA	65	30	62	56	32	43	57	63	42	37	258	Flowering plant	147
37	TSETSE FLY	65	31	74	42	19	63	34	76	65	24	257	Invertebrate	138
38	SHARK	75	23	27	77	13	68	65	51	76	14	256	Fish	71
39	COTTON	30	69	86	24	92	5	42	72	5	93	255	Flowering plant	257
40	BAMBOO	89	15	68	50	21	61	34	77	39	40	251	Flowering plant	153
41	ARCHAEOPTERYX	92	11	13	87	45	29	80	19	21	73	251	Bird	121
42	DOG	40	58	96	2	24	51	80	20	9	82	249	Mammal	336
43	POTATO	50	48	89	20	21	62	62	59	24	68	246	Flowering plant	210
44	ELEPHANT	20	77	76	34	76	13	42	73	32	57	246	Mammal	175
45	TICK	60	38	65	53	13	69	45	67	62	27	245	Invertebrate	415
46	CAT	45	51	88	22	24	52	79	23	8	85	244	Mammal	340
47	HERPES	56	39	84	27	13	70	78	27	12	81	243	Virus	422
48	PIG	35	65	77	33	56	23	68	45	7	88	243	Mammal	221
49	QUETZALCOATLUS	45	52	13	88	30	48	90	7	63	25	241	Reptile	118
50	DRAGONFLY	65	32	13	89	13	71	76	30	72	17	239	Invertebrate	101

What on Earth Evolved? . . . in Brief

OVERALL RANK	SPECIES	EVOLUTIONARY IMPACT		IMPACT ON HUMAN HISTORY		ENVIRONMENTAL IMPACT		GLOBAL REACH		LONGEVITY		TOTAL SCORE	GROUP	PAGE NUMBER
		SCORE	RANK	SCORE	RANK	SCORE	RANK	SCORE	RANK	SCORE	RANK			
51	LOBE-FINNED FISH	94	6	13	90	34	36	23	91	72	18	236	Fish	106
52	PROTOTAXITES	94	7	13	91	34	37	70	42	24	69	235	Fungus	76
53	APPLE	22	76	76	35	33	41	72	39	31	58	234	Flowering plant	358
54	CANNABIS	45	53	64	55	13	72	75	35	37	42	234	Flowering plant	292
55	DEER	52	44	45	63	34	38	76	31	25	65	232	Mammal	396
56	VELVET WORM	95	4	13	92	13	73	31	82	80	10	232	Invertebrate	62
57	COD	35	66	68	51	42	33	65	52	20	75	230	Fish	218
58	HORSE	38	61	94	6	13	74	76	32	7	89	228	Mammal	246
59	SPERM WHALE	20	78	57	60	23	56	87	8	40	39	227	Mammal	171
60	AZOLLA	13	88	18	80	90	6	44	68	61	30	226	Non-flowering plant	86
61	CHICKEN	35	67	75	38	32	44	76	33	6	92	224	Bird	232
62	ROSE	20	79	58	58	45	30	63	57	37	43	223	Flowering plant	354
63	SMALLPOX	19	87	94	7	18	64	80	21	9	83	220	Virus	13
64	CROW	55	42	27	78	23	57	80	22	31	59	216	Bird	405
65	RED FOX	45	54	42	67	23	58	81	16	25	66	216	Mammal	400
66	LYSTROSAURUS	84	20	13	93	24	53	82	15	13	79	216	Reptile	115
67	TOBACCO	20	80	91	15	32	45	32	80	31	60	206	Flowering plant	297
68	DUNG BEETLE	13	89	43	66	32	46	54	65	62	28	204	Invertebrate	97
69	ERGOT	34	68	54	61	13	75	65	53	37	44	203	Fungus	317
70	ANTHRAX	13	90	18	81	13	76	66	50	90	3	200	Bacterium	23
71	TEA	13	91	76	36	44	32	29	84	37	45	199	Flowering plant	287
72	SPONGE	20	81	15	82	13	77	60	61	90	4	198	Invertebrate	48
73	GRAPE	13	92	84	28	13	78	58	62	28	61	196	Flowering plant	307
74	SLIME MOULD	65	33	13	94	13	79	27	86	78	12	196	Protoctist	36
75	BAT	55	43	26	79	13	80	65	54	35	51	194	Mammal	168

The Top 100 Species

OVERALL RANK	SPECIES	EVOLUTIONARY IMPACT		IMPACT ON HUMAN HISTORY		ENVIRONMENTAL IMPACT		GLOBAL REACH		LONGEVITY		TOTAL SCORE	GROUP	PAGE NUMBER
		SCORE	RANK	SCORE	RANK	SCORE	RANK	SCORE	RANK	SCORE	RANK			
76	RUBBER TREE	25	72	72	45	34	39	27	87	34	52	192	Flowering plant	261
77	HIV/AIDS	20	82	76	37	13	81	67	46	13	80	189	Virus	7
78	SEA SCORPION	52	45	13	95	23	59	55	64	44	35	187	Invertebrate	65
79	POPPY	20	83	72	46	13	82	46	66	33	55	184	Flowering plant	321
80	OLIVE	40	59	72	47	13	83	27	88	28	62	180	Flowering plant	214
81	CACAO	25	73	71	48	23	60	34	78	27	64	180	Flowering plant	278
82	DIMETRODON	89	16	13	96	13	84	35	74	23	72	173	Reptile	112
83	RHYNIOPHYTE	92	12	13	97	13	85	35	75	20	76	173	Non-flowering plant	79
84	CINCHONA	24	75	78	32	13	86	26	90	28	63	169	Flowering plant	330
85	COFFEE	13	93	54	62	48	27	28	85	24	70	167	Flowering plant	283
86	SILKWORM	20	84	75	39	13	87	43	70	7	90	158	Invertebrate	266
87	CAMEL	36	62	74	43	13	88	27	89	5	94	155	Mammal	252
88	TIKTAALIK	93	10	13	98	13	89	13	95	18	77	150	Amphibian	109
89	DURIAN	62	35	13	99	13	90	13	96	48	31	149	Flowering plant	150
90	HAMSTER	20	85	34	71	13	91	76	34	1	100	144	Mammal	349
91	BLACK PEPPER	13	94	62	57	13	92	19	92	37	46	144	Flowering plant	370
92	CHILLI PEPPER	39	60	29	76	13	93	13	97	37	47	131	Flowering plant	377
93	LAVENDER	13	95	32	72	13	94	32	81	37	48	127	Flowering plant	366
94	PSEUDOMONAS	13	96	31	75	66	18	13	98	3	97	126	Bacterium	26
95	ORANGE	25	74	39	69	13	95	44	69	4	96	125	Flowering plant	386
96	COCA	13	97	32	73	13	96	18	94	37	49	113	Flowering plant	312
97	VANILLA	13	98	36	70	13	97	13	99	37	50	112	Flowering plant	362
98	BANANA	20	86	44	64	13	98	31	83	3	98	111	Flowering plant	390
99	LOTUS	13	99	32	74	13	99	19	93	33	56	110	Flowering plant	374
100	POTYVIRUS	13	100	58	59	13	100	13	100	2	99	99	Virus	10

Postscript

Thirty species that nearly made it!

PAINFUL DECISIONS ABOUT what to include or leave out in the top 100 have inevitably had to be made. Here I console myself with a list of those I would like to have featured but could not because of overlaps, repetition or for fear of creating a list that was insufficiently representative of the diversity of species on Earth. Please feel free to comment, debate, agree or disagree with any of my choices on the forum at the website: www.whatonearthevolved.com. All contributions are welcomed, responded to and gratefully received.

	Common Name	Species (unless stated)	Summary
1	COCCOLITHOPHORE	*Emiliania huxleyi*	One of the most widely distributed single-celled micro-organisms in the seas, it is thought to have had a significant impact on the Earth's biosphere.
2	CYCAD	*Cycadales* (Order)	The first plants to develop seeds as a way of protecting their embryos.
3	SPIDER	*Parasteatoda tepidariorum*	Evolving approximately 380 million years ago from a crab/scorpion-like ancestor, there are about 40,000 spider species alive today. *Parasteatoda tepidariorum* is especially common in America.

What on Earth Evolved? . . . in Brief

	Common Name	Species (unless stated)	Summary
4	EDAPHOSAURUS	*Edaphosaurus boanerges*	A sail-back pelycosaur from the same family as *Dimetrodon* (see page 112) but with a significant claim to fame as being one of the first (if not the first) herbivorous land animals.
5	ICHTHYOSAURUS	*Ichthyosaurus breviceps*	An extinct marine reptile and one of the first creatures able to give birth to live young (viviparous).
6	CROCODILE	*Crocodylus acutus*	Large semi-aquatic reptile that somehow survived the extinction event that led to the death of the non-avian dinosaurs 65.5 million years ago.
7	COMMON DAISY	*Bellis perennis*	One of the few non-cultivated flowers (or weeds) to be admired for their bright colours and perfect symmetry (see also page 374).
8	TERMITE	*Mastotermes darwiniensis*	Descended from cockroaches, these small highly social wood-eating insects build enormous nests, sometimes taller than a house, usually containing several million individuals.
9	NEANDERTHAL	*Homo neanderthalensis*	An extinct member of the human genus that evolved approximately 500,000 years ago, with the last individuals dying out in Europe about 25,000 years ago.
10	GOLDEN EAGLE	*Aquila chrysaetos*	Well-known birds of prey that have been instrumental in controlling populations of insects, mammals and reptiles, including mice, rats and snakes.
11	SNAKE	*Titanoboa cerrejonensis*	Carnivorous reptiles descended from lizards that lost their legs, adapting to life on land in a manner that lets them coil up in trees and under long grasses. Titanoboa can grow up to thirteen metres long!

Thirty Species That Nearly Made It!

	Common Name	Species (unless stated)	Summary
12	GREY WOLF	*Canis lupus*	Wolves can survive in almost any habitat, including forests, deserts, mountains and urban areas.
13	GORILLA	*Gorilla gorilla*	Vegetarian, knuckle-walking primates that survive in only two African species but have an extremely close genetic relationship with humans.
14	SUGAR BEET	*Beta vulgaris*	Sugar beet now accounts for 30 per cent of the world's sugar production, so although the little brother of sugarcane (see page 206), it plays a significant part in modern agriculture.
15	GOAT	*Capra aegagrus*	One of the first animal species to be domesticated in south-west Asia and eastern Europe. Being such a close cousin of the sheep precluded it from the top 100 species.
16	MOUSE	*Mus musculus domesticus*	As the most common laboratory animals, mice are key to the search for cures for human diseases.
17	SOYA BEAN	*Glycine max*	The major impact of this crop on modern human history brought it to within a whisker of inclusion in the top 100. Its exclusion was mainly due to its relatively recent rise to global status.
18	FLAX	*Linum usitatissimum*	An extremely versatile crop grown since ancient times around the Mediterranean and the Middle East for its seeds, fibres and oil. Flax is not grown in such large quantities today, which is why it only nearly made it.

	Common Name	Species (unless stated)	Summary
19	FROG	Anura (Order)	Frogs are one of the most ancient and diverse orders of amphibians (about 5,000 species are known, and fossils date back 250 million years) and have adapted to live in regions ranging from the tropics to the Sub-Arctic.
20	KOLA NUTS	*Cola* (Genus)	The other principal constituent in the world's 'favourite drink' (see page 314).
21	PLASMODIUM	*Plasmodium falciparum*	These microscopic parasitic protozoa which cause malaria have already featured in the story of the mosquito (see page 131) and cinchona (see page 331).
22	TYPHOID	*Salmonella typhi*	A disease-causing bacterium that spreads by contaminated food and water, or is carried on the legs of flying insects (see fruit fly, page 413), and survives phagocytosis by the human immune system (see slime mould, page 37).
23	CHOLERA	*Vibrio cholerae*	Historically this bacterium has been a significant human killer, causing death by dehydration from acute diarrhoea. If left untreated mortality rates can reach as high as 60 per cent.
24	TUBERCULOSIS	*Mycobacterium tuberculosis*	A bacterial disease that attacks the lungs and other parts of the body. It is established in up to a third of the world's population – mostly in Asia and Africa – although only a minority develop the full-blown disease.

Thirty Species That Nearly Made It!

	Common Name	Species (unless stated)	Summary
25	POLIO	*Poliovirus*	The virus that causes the disease poliomyelitis, resulting in wasted limbs, paralysis and sometimes death. A vaccine was pioneered by American virologist Jonas Salk in 1952 and mass vaccination programmes began five years later.
26	SUNFLOWER	*Helianthus annuus*	Another asterid (see also common daisy, 442) but with a much larger flowering head and seeds that can be turned into a valuable source of nutrition, such as cooking oil and margarine, as well as bio-diesel.
27	RUST	*Puccinia triticina*	A fungal parasite that causes widespread economic damage to crops such as wheat and rye.
28	LEMON	*Citrus limon*	Genetically so similar to oranges (see page 386) that to include it in the top 100 would have been over-generous to the genus *Citrus*.
29	TOMATO	*Solanum lycopersicum*	Native to South America, yellow tomato varieties were the first to be domesticated (by *c.* 500 BC). They began to be grown in Spain, Italy and England from the 1600s. More than one hundred million tonnes of tomatoes are harvested annually today.
30	PINWORM	*Enterobius vermicularis*	A type of roundworm (see page 52) which has thrived with the growth of human populations and infects an estimated 200 million people worldwide.

Further Reading

If you'd like to read the endnotes for this book, they can be found at www.whatonearthevolved.com.

General

On the Origin of Species by Charles Darwin (John Murray, 1859)

The Descent of Man by Charles Darwin (John Murray, 1871)

Principles of Geology by Charles Lyell (John Murray, 1872)

Genome by Matt Ridley (Fourth Estate, 1999)

Diversity of Life by Lynn Margulis, Karlene Schwartz and Michael Dolan (Jones and Bartlett, 1999)

The Selfish Gene by Richard Dawkins (Oxford University Press, 1976)

The Crucible of Creation by Simon Conway Morris (Oxford University Press, 1998)

The Variety of Life by Colin Tudge (Oxford University Press, 2002)

The Rise of Life by John Reader (Collins, 1986)

The Complete Guide to Prehistoric Life by Tim Haines and Paul Chambers (BBC Books, 2003)

The Ancestor's Tale by Richard Dawkins (Houghton Mifflin, 2004)

A Natural History of Domesticated Animals by Juliet Clutton-Brock (Cambridge University Press, 1999)

1. On Viruses

Viruses and the Evolution of Life by Luis Villarreal (ASM, 2005)

Princes and Peasants by Donald Hopkins (University of Chicago Press, 1983)

The Life and Death of Smallpox by Ian and Jenifer Glynn (Profile, 2004)

The Tulip by Anna Pavord (Bloomsbury, 1999)

2. On Simple Cells

The Origins of Life by John Maynard Smith and Eörs Szathmáry (Oxford University Press, 2000)

Enriching the Earth by Vaclac Smil (MIT Press, 2004)

Oxygen: The Molecule that Made the World by Nick Lane (Oxford University Press, 2002)

The Ages of Gaia by James Lovelock (Oxford University Press, 1988)

Acquiring Genomes: A Theory of the Origins of Species by Lynn Margulis and Dorion Sagan (Basic Books, 2002)

The World Without Us by Alan Wiseman (Viking, 2007)

3. On Symbiosis

Emergence: The Connected Lives of Ants, Brains, Cities and Software by Steven Johnson (Allen Lane, 2001)

The Red Queen by Matt Ridley (Viking, 1993)

Propitious Esculent by John Reader (Heinemann, 2008)

4. On Sea Life

Trilobite! by Richard Fortey (Flamingo, 2001)

The Theory of Evolution by J. M. Smith (Cassell, 1962)

Acquiring Genomes by Lynn Margulis and Dorion Sagan (Basic Books, 2002)

In the Blink of an Eye: How Vision Kick-started the Big Bang of Evolution by Andrew Parker (Free Press, 2004)

Old Fourlegs by J. L. B. Smith (Longman, 1938)

Endless Forms Most Beautiful by Sean Carroll (Weidenfeld & Nicolson, 2006)

5. On Pioneers of the Land

Why Size Matters by John Bonner (Princeton University Press, 2006)

The Secret Lives of Plants by David Attenborough (BBC Books, 1995)

The Earth Moved by Amy Stewart (Frances Lincoln, 2004)

Fungi by Roy Watling (Natural History Museum, 2003)

Further Reading

Amber by Andrew Ross (Natural History Museum, 1998)

Evolution of Insects by David Grimaldi and Michael Engels (Cambridge University Press, 2005)

Fossil Plants by Paul Kenrick and Paul Davis (Natural History Museum, 2004)

6. On Fish that Came Ashore

Your Inner Fish by Neil Shubin (Allen Lane, 2008)

The Emerald Planet by David Beerling (Oxford University Press, 2007)

Extinctions in the History of Life edited by Paul Taylor (Cambridge University Press, 2004)

7. On Biodiversity

The Insect Societies by E. O. Wilson (Harvard University Press, 1974)

Fleas, Flukes and Cuckoos by Miriam Rothschild and Theresa Clay (Collins, 1952)

The Diversity of Life by E. O. Wilson (Allen Lane, 1993)

8. On the Rise of Reason

The Origins of Virtue by Matt Ridley (Viking, 1996)

Bats in Question by D. E. Wilson (Smithsonian Institution Press, 1997)

The Mating Mind by Geoffrey Miller (Heinemann, 2000)

Leviathan: The History of Whaling in America by Eric Dolin (Norton, 2007)

Coming of Age with Elephants by Joyce Poole (Hodder & Stoughton, 1996)

The Elephant's Secret Sense by Caitlin O'Connell (Free Press, 2007)

The Story of Rats by S. Anthony Barnett (Allen & Unwin, 2001)

After Man: A Zoology of the Future by Dougal Dixon (Granada, 1981)

9. On Agriculture

Food, A History by Felipe Fernández-Armesto (Macmillan, 2001)

Plants, Man and Life by Edgar Anderson (Andrew Melrose, 1954)

Sugar: The Grass that Changed the World by Sanjida O'Connell (Virgin, 2004)

The Potato by Larry Zuckerman (Macmillan, 1999)

A History of the British Pig by Julian Wiseman (Duckworth, 1986)

Sheep by Alan Butler (O Books, 2006)

Cattle by Valeria Porter (The Crowood Press, 2007)

China: Land of Discovery and Invention by Robert Temple (Patrick Stephens, 1986)

People, Plants and Genes by Denis Murphy (Oxford University Press, 2007)

Planet Chicken by Hattie Ellis (Sceptre, 2007)

Cod: A Biography of the Fish that Changed the World by Mark Kurlansky (Jonathan Cape, 1998)

10. On Material Wealth

War Horse by Louis Dimarco (Westholme Yardley, 2008)

Inside Your Horse's Mind by Lesley Skipper (J. A. Allen, 1999)

The Behaviour of Horses by Marthe Kiley-Worthington (J. A. Allen, 1987)

The Camel and the Wheel by Richard Bulliet (Harvard University Press, 1975)

The Camel in Australia by Tom McKnight (Melbourne University Press, 1969)

The Fibre that Changed the World edited by Douglas Farnie (Oxford University Press, 2004)

Tears of the Tree by John Loadman (Oxford University Press, 2005)

Jacquard's Web by James Essinger (Oxford University Press, 2004)

11. On Drugs

On Drugs by David Lenson (University of Minnesota Press, 1995)

Tastes of Paradise by Wolfgang Schivelbusch (Vintage, 1992)

Matters of Substance by Griffin Edwards (Allen Lane, 2004)

Consuming Habits edited by Jordan Goodman, Paul E. Lovejoy and Andrew Sherratt (Routledge, 1995)

The Pursuit of Pleasure by Rudi Matthee (Princeton University Press, 2005)

Forces of Habit by David Courtwright (Harvard University Press, 2001)

Fungi by Brian Spooner and Peter Roberts (Collins, 2005)

A Short History of Wine by Rod Phillips (Allen Lane, 2000)

Phylloxera: How Wine was Saved for the World by Christy Campbell (HarperPerennial, 2004)

Further Reading

Narcotic Culture: A History of Drugs in China by Frank Dikötter et al. (Hurst & Company, 2004)

The Alchemy of Culture by Richard Rudgeley (British Museum Press, 1998)

12. On Companionship

Man and the Natural World by Keith Thomas (Allen Lane, 1983)

Dog by Susan McHugh (Reaktion Books, 2004)

Cat by Katherine Rogers (Reaktion Books, 2006)

Rabbits and their History by John Sheail (David & Charles, 1971)

Hamsters & Guinea Pigs by David le Roi (Nicholas Vane, 1955)

13. On Beauty

A History of the Fragrant Rose by Allen Patterson (Little Books, 2007)

The Botany of Desire by Michael Pollan (Bloomsbury, 2002)

Vanilla by Tim Ecott (Penguin, 2004)

Lavender (The Genus Lavandula) by Maria Lis-Balchin (CRC Press, 2002)

Out of the East by Paul Freedman (Yale University Press, 2008)

The Encyclopedia of Grasses for Livable Landscapes by Rick Darke (Timber Press, 2007)

The Grass is Greener by Tom Fort (HarperCollins, 2000)

The Citrus Industry by Herbert John Webber et al. (University of California, 1967)

Curry by Lizzie Collingham (Chatto & Windus, 2005)

Bananas: An American History by Virginia Scott Jenkins (Smithsonian Institution Press, 2000)

14. On Rivalry

Something New Under the Sun by J. R. McNeill (Allen Lane, 2000)

Canids by Claudio Sillero-Zubiri, Michael Hoffmann and David W. Macdonald (Union Internationale pour la Conservation de la Nature et de ses Ressources, 2004)

In the Company of Crows and Ravens by John Marzluff and Tony Angell (Yale University, 2005)

After Man: A Zoology of the Future by Dougal Dixon (Granada, 1981)

What on Earth Evolved? . . . in Brief

Time, Love, Memory by Jonathan Weiner (Faber & Faber, 1999)

Fly by Steve Connor (Reaktion Books, 2006)

Maggots, Murder and Men by Zakaria Erzinclioglu (Colchester Books, 2000)

Deer of the World by G. Kenneth Whitehard (Constable, 1972)

Index

Index

and drugs/addictive substances, 288–9, 300, 301, 305, 320, 323, 325
and material wealth, 257–9, 264, 269
see also England; Scotland
British Museum, 134
Broeck, Pieter van den, 285
Bronze Age, 83
Brown, Lancelot 'Capability', 382
Buddha, 207, 374–5
Buddhism, 15, 207, 229, 316
buffalo, 229–30
Bulgaria, 355
Burgdorfer, Willy, 415
Burgess Shale, 58, 93
Burgessochaeta, 93
Burma, 157, 324
Byzantine Empire, 268

C

cacao, 278–82, 434, 439
Cadbury's, 281
Caenorhabditis elegans, 56, 57
Caesar, Julius, 367
caffeine, 208, 283, 290
Cairo, 252, 254, 284
calcium carbonate, 46, 52
Calculating Machine, 269
Calicut, 372
California, 217, 239, 255, 273, 386, 389
Cambrian Period, 53, 59, 60, 62, 65, 68, 93
camel, 172, 252–6, 432, 439
    Bactrian, 253
    dromedaries, 253
Canada, 42, 58, 110, 125, 417
Canary Islands, 230, 390
cannabis, 292–6, 438
cannibalism, 241, 242
Canton, 289
Cape of Good Hope, 140
Cape Town, 7
capsaicin, 377–8
carbon dioxide, 45, 47, 53, 70, 81, 84, 87, 88, 115, 116, 303
Carboniferous Period, 83, 102, 103, 430
Carcharodon megalodon, 71–3
Caribbean, 208, 286, 390
Carson, Rachel, 95
    Silent Spring, 95
Carter, Howard, 367
Casement, Sir Roger, 263
Castilla elastica, 261
cat, 340–43, 431, 437
Catherine de Medici, 367–8
Catholic Church/Catholicism, 42, 212
cattle *see* cow/cattle
cave paintings, 194, 405
cells, 37–8, 48–9
    single *see* Prokaryotes
Central Africa, 140
    see also Africa
Central America, 15, 31, 41, 147, 241, 261, 279, 282, 390, 391, 393 *see also* Americas, the
Cervidae, 396, 398
    see also deer

Ceryneian Hind, 397
Ceylon (now Sri Lanka), 264, 332
Chaco Canyon, 242
Chain, Ernst, 327
chakras, 375
Champagne, 226
Chandragupta Maurya, 177
Chapman, John ('Johnny Appleseed'), 358–60
    Charles I, King, 367
Charles II, King of England, 331
Charles IX, King of France, 367
Charleston, 239
Cheyenne River Indian Reservation, 126
Chicago: Field Museum, 125–6
    University of, 76
chicken, 232–5, 402, 438
Chile, 30, 32, 233, 311
chilli pepper, 377–80, 439
chimpanzees, 184
China, 15, 31, 32, 122, 136, 137, 159, 163, 187, 238, 398, 400
    and agriculture, 203, 207, 210, 216, 221, 223, 225, 229–30, 233–4, 236–7, 242
    and beauty (looks/smells/flavours), 356, 358, 381, 388, 389
    and drugs/addictive substances, 287–8, 289, 290, 294, 297, 299, 316, 323–4
    and material wealth, 251, 258, 267–8, 274
Chlamydomonas reinhardtii, 47
chocolate, 208, 278–82
cholera, 444
chordates, 60
Christians/Christianity, 15, 144, 182, 203, 207–8, 268, 309, 330
cider, 360
cigarettes, 298–301
cinchona, 330–33, 434, 439
Circa Instans, 371
Cistercians, 145, 225, 226
citrus fruits, 386, 387
    see also lemon; orange
clades, 106
Cleopatra, 367
climate, 54, 70, 83, 84, 86, 92, 126, 137, 168, 188, 415
clouds, 46
Clovis I, 397
club mosses, 82, 83, 90
    see also Lepidodendron
Clusius, Carolus, 11
Cnidaria, 52, 56, 60
coal, 84, 91, 116
coca, 312–16, 434, 439
Coca-Cola, 312, 314, 315
cocaine, 312, 313–16
    'crack', 315–16
coccolithophores, 45, 441
cockfighting, 232, 233
cod, 218–20
codeine, 321
cod wars, 220
coffee, 283–6, 439
collaboration, 37, 52
collagen, 50
Colombia, 43, 312, 349

Index

What on Earth Evolved? . . . in Brief

Index

Index

Index

Acknowledgements

EVERY LIVING THING that I have encountered whilst writing this book has been a unique privilege and joy to research, interpret and describe. What extraordinary stories they all have to tell! Ultimately, therefore, it is only thanks to nature's unfathomable genetic diversity that such a survey has been possible.

However, discovering and weaving such tales into a connected narrative would have been completely impossible were it not for the scholarship of literally thousands of people – too numerous to mention by name. Their many books, articles and conversations have been integral to my attempt to turn a patch-work quilt into what is hopefully a meaningful, informative and interconnected illumination of the story of life on Earth and the place of humanity within it.

When considering the many people who have worked tirelessly to turn this text into such a beautifully illustrated book, things get personal. **Richard Atkinson**, my superlative commissioning editor, has been a tower of support and inspiration from start to finish. **Natalie Hunt**, editor-in-chief of the *What on Earth . . . ?* series, has worked with unbounded professionalism, often seven days a week, to ensure that despite our ambitious deadlines this project has been safely delivered without missing a beat. Meanwhile, illustrative genius **Andy Forshaw** has yet again sprinkled his magic pictures on just about every page while **Will Webb**, designer, has laid out each spread with total skill, love and care. Thanks also to **Richard Emerson** who scrutinized the text, making many useful comments, suggestions and corrections throughout and to **Andrew Lownie**, my literary agent, for his ongoing advice, encouragement and support.

My deep gratitude also goes to blue-badge tour-guide **Anne Marie Ehrlich** (on pictures), Bond-girl **Penelope Beech** (on marketing), cockroach-hating **Anya Rosenberg** (on publicity) and Turkish delights **Ruth Logan** and **Katie Mitchell** (on rights – responsible for masterminding the translation of *What on Earth Happened?* into fourteen languages so far . . .). Thanks also to **Polly Napper** (on production), **Sarah Barlow** (on final polish), expert mathematician **Erica Jarnes** (on fact-checking), sales supremo **David Ward** (on sales), **Trâm-Anh Doan** (for the paperback edition) and the entire Bloomsbury sales force (including the very tall and super-diligent **Andrew Sauerwine**). Finally, my sincere thanks to **Richard Charkin** (on management) for his continued faith in and support for the *What on Earth . . . ?* concept.

I dedicate this book to my wonderful wife **Gins**, our two lovely daughters **Matilda** and **Verity** and also to little **Flossie**, who has rightly come out top dog (see page 336). It is entirely thanks to your constant inspiration, love, patience, company and support that my relentless wrestling with all this stuff about the planet, life and people has been in the least bit possible.

Comments and suggestions about this boom can be posted on the discussion forum at **www.whatonearthevolved.com**. All contributions and feedback are most welcome.